LEVEL
B

SRA
Connecting
Math Concepts

SRA

Columbus, Ohio

The **McGraw·Hill** Companies

TABLE OF CONTENTS

Instructions

This book contains blackline masters that can be used with *Connecting Math Concepts* Level B. Math facts are available for every lesson from 21 through 120. The number on the sheet indicates the earliest lesson on which a sheet is to be presented; however, sheets may be repeated and sheets may be skipped.

All sheets have a title and the title is coordinated with the sequence of facts in *Connecting Math Concepts,* Level B.

Some sheets have stars (*) at the top. Those are introductory sheets. Starred sheets are to be preceded by a verbal presentation. Other sheets have a **T** at the top. Those are test sheets.

As a rule, present the various starred sheets and the following test sheets. There may be other sheets between the starred and test sheets. Use those sheets if necessary to firm the children. Note: It is not necessary to present all the sheets to every group of children. Criterion: If children do well on the starred sheets for a particular series (such as +2 or –9), present the test sheet on the next lesson. If children do not do well, present intermediate sheets before presenting the test sheet.

If children do not pass the test sheet, present one or more of the preceding practice sheets, then repeat the test sheet.

Procedure for starred (*) sheets:

Verbally present the first 12 problems on the sheet before presenting the sheet. Make sure that children are responding quickly and accurately. Time children on the worksheet. Check answers. Provide remedies based on the children's performance.

1. Verbal work. "I'm going to say some problems. You'll tell me the answer."
   Present the first three rows of problems on the sheet.
   Say each problem as follows:
   Example: "One plus one. What's the answer?"
   "One plus zero. What's the answer?", etc.
   Correct mistakes and repeat the first three rows until the children are responding on signal and correctly.

2. Present the worksheet. "Pencils down. Don't write anything until I say 'Go.' You'll do the problems a row at a time. Do not do the problems a column at a time. You have **two minutes** to complete the sheet. Get ready. Go."
   (Observe children and give feedback. Children are to start with the upper left problem and work across the rows, not down the columns. Children are not to count on their fingers or take an inordinate amount of time in "figuring out" the answer.)
   At the end of two minutes. "Stop. Pencils down."

3. "Check your answers. I'll say each fact starting with the first row. Make an X next to any fact you missed."
   "One plus one equals two."
   "One plus zero equals one," etc.
   Children are to mark every problem they did not complete as a problem that is missed.
   "Write the number of problems you missed at the top of the sheet."
   (Observe children and give feedback.)
   "Write the correct fact above any item you missed. Don't cross out or erase what you wrote."

Children who made no more than 3 errors do not have to repeat the sheet. Children who made 4 or more errors must repeat the sheet on their own. (This work will not be timed.)
Note: For lesson 96 and 97, read 9+[ ]=11 as "9 plus what number equals 11?"

PROCEDURE FOR SHEETS WITH NO STAR OR T:

Do not precede the sheet with a verbal presentation.
Present the worksheet and time it, and check it, as above.

PROCEDURE FOR SHEETS WITH A T (TEST SHEETS):

Do not precede the sheet with a verbal presentation.
Present the worksheet and time it. Check it, as above.
Record the performance of each child on the record page.
Write the number of errors each child made.
Children who make no more than 3 errors pass the Test. Children who make more than 3 errors must repeat the sheet and be timed on the sheet. When these children pass the sheet, write and circle the number of errors they made when they passed.

Connecting Math Concepts, Level B
Math Facts Worksheets (Lessons 21-120)

Topic	Lessons
Plus 0s and 1s	21*, 22, 23*, 24-30, 37, 40
Minus 0s and 1s	31*, 32, 36, 38
Plus/Minus 0s and 1s	33*, 34-35, 39
Plus 2s	41*, 42, 43*, 44-46, 47*, 48-50
Minus 2s	51*, 52
Plus Mix: 0s, 1s, 2s	53, 54, 58, 59
Minus Mix: 0s, 1s, 2s	55-57, 60
Plus 10s	61*, 62-64, 72
Plus Mix: 0s, 1s, 2s, 10s	65-69, 73
Minus Mix: 0s, 1s, 2s	70, 71, 74
Plus 9s	75*, 76*, 77, 78*, 79, 80, 90, 98, 103, 118
Minus 10s	81*
Minus Mix: 0s, 1s, 2s, 10s	82*, 83, 84
Plus 3s	85*, 86*, 87, 88*, 89
Plus Mix: 1s, 2s, 3s	89, 95, 108
Minus 9s	91*, 92*, 93, 94, 100, 101
9 Plus Algebra	96*, 97*, 102, 104, 109, 115, 119
Plus Doubles	105*, 106*, 107, 110, 117
Minus Mix: 1s, 2s, 9s, 10s	111*, 112
2-Digit-2 Plus	113*, 114, 116, 120

Key for sheets marked * or [T]

* **Introductory Sheets.** Follow verbal presentation procedure for these lessons: 21, 23, 31, 33, 41, 43, 47, 51, 61, 75, 76, 78, 81, 82, 85, 86, 88, 91, 92, 96, 97, 105, 106, 111, 113.

[T] **Test Sheets.** Follow test sheet procedures for these lessons: 30, 32, 39, 50, 52, 54, 57, 64, 69, 80, 84, 87, 94, 110, 112, 116.

Math Facts Record Sheet Form, Level B

Record number of errors each child made. Children who make more than 3 errors must repeat the sheet.

Name	30T	32T	39T	50T	52T	54T	57T	64T	69T	80T	84T	87T	94T	110T	112T	116T

Lesson 21*   Plus 0s and 1s

Facts: 1+, 0+: a

Name: _______________________

1+1=	1+0=	1+2=	1+4=
1+3=	1+5=	1+1=	1+5=
1+4=	1+1=	1+2=	0+3=
1+2=	1+3=	1+4=	1+2=
1+1=	1+0=	1+5=	1+4=
1+5=	1+2=	0+1=	1+0=
1+3=	1+0=	1+2=	1+3=
1+2=	1+5=	1+4=	1+5=
1+3=	1+4=	1+0=	1+1=
1+5=	1+3=	1+1=	1+2=
0+4=	1+1=	1+5=	1+0=

Lesson 22   Plus 0s and 1s

Name: _______________________

1+3=	1+4=	1+5=	1+1=
1+2=	1+0=	1+1=	1+3=
1+0=	1+3=	0+1=	1+0=
0+5=	1+2=	1+1=	0+4=
1+2=	1+5=	1+0=	1+2=
1+4=	1+3=	1+1=	1+3=
1+2=	1+1=	0+3=	1+0=
1+3=	1+4=	1+2=	1+1=
1+2=	1+3=	1+4=	1+5=
1+1=	1+0=	1+2=	1+3=
1+5=	1+4=	1+0=	0+2=

Name: _______________________________

9+1=	4+1=	6+1=	0+5=
3+0=	5+1=	3+1=	7+0=
1+1=	7+1=	10+1=	1+0=
1+0=	5+0=	2+0=	6+1=
4+1=	7+0=	8+0=	3+1=
5+1=	1+0=	9+1=	10+1=
7+0=	6+1=	5+1=	2+0=
6+1=	3+1=	1+0=	7+0=
3+1=	0+4=	4+1=	8+0=
10+1=	2+0=	5+1=	9+1=
2+0=	8+0=	7+0=	8+0=

Lesson 24   Plus 0s and 1s

Name: _______________________

5+1=	4+1=	1+1=	15+1=
7+0=	0+3=	2+1=	17+0=
6+1=	7+1=	10+1=	11+0=
9+1=	5+1=	3+0=	16+1=
3+0=	7+0=	6+0=	13+1=
1+1=	1+0=	8+1=	10+1=
1+0=	6+1=	7+1=	12+0=
4+1=	3+1=	1+1=	17+0=
10+1=	4+1=	4+0=	18+0=
2+0=	0+10=	4+1=	9+1=
1+1=	8+0=	9+1=	18+0=

Lesson 25   Plus 0s and 1s

Name: ________________________________

7+1=	20+1=	1+4=	1+5=
4+1=	50+1=	0+12=	1+11=
9+1=	60+1=	1+6=	1+8=
3+1=	30+1=	1+9=	1+4=
6+0=	90+1=	1+3=	1+1=
8+1=	40+1=	1+7=	1+16=
2+1=	3+0=	1+4=	1+5=
11+1=	7+1=	1+6=	1+11=
1+1=	4+1=	0+10=	1+9=
1+16=	9+1=	50+1=	1+4=
30+1=	1+3=	1+16=	1+0=

Name: _______________________

1+5=	6+1=	60+1=	6+1=
1+11=	3+0=	30+1=	8+1=
1+8=	8+1=	90+1=	2+1=
1+4=	1+3=	40+0=	11+1=
1+1=	40+1=	1+15=	15+1=
8+1=	12+1=	1+80=	1+16=
15+1=	0+5=	1+40=	7+1=
1+9=	1+11=	1+50=	4+1=
8+1=	1+6=	1+18=	9+1=
1+9=	1+4=	1+8=	3+1=
1+1=	12+0=	1+50=	6+1=

Lesson 27   Plus 0s and 1s

Name: _______________________________

9+1=	4+1=	6+1=	0+5=
3+0=	5+1=	3+1=	7+0=
1+1=	7+1=	10+1=	1+0=
1+0=	5+0=	2+0=	6+1=
4+1=	7+0=	8+0=	3+1=
5+1=	1+0=	9+1=	10+1=
7+0=	6+1=	5+1=	2+0=
6+1=	3+1=	1+0=	7+0=
3+1=	0+4=	4+1=	8+0=
10+1=	2+0=	5+1=	9+1=
2+0=	8+0=	7+0=	8+0=

Lesson 28   Plus 0s and 1s

Name: _______________________________

7+1=	20+1=	1+4=	1+5=
4+1=	50+1=	0+12=	1+11=
9+1=	60+1=	1+6=	1+8=
3+1=	30+1=	1+9=	1+4=
6+0=	90+1=	1+3=	1+1=
8+1=	40+1=	1+7=	1+16=
2+1=	3+0=	1+4=	1+5=
11+1=	7+1=	1+6=	1+11=
1+1=	4+1=	0+10=	1+9=
1+16=	9+1=	50+1=	1+4=
30+1=	1+3=	1+16=	1+0=

Lesson 29   Plus 0s and 1s

Name: _______________________________

5+1=	4+1=	1+1=	15+1=
7+0=	0+3=	2+1=	17+0=
6+1=	7+1=	10+1=	11+0=
9+1=	5+1=	3+0=	16+1=
3+0=	7+0=	6+0=	13+1=
1+1=	1+0=	8+1=	10+1=
1+0=	6+1=	7+1=	12+0=
4+1=	3+1=	1+1=	17+0=
10+1=	4+1=	4+0=	18+0=
2+0=	0+10=	4+1=	9+1=
1+1=	8+0=	9+1=	18+0=

Lesson 30 T   Plus 0s and 1s

Name: _______________________________

1+5=	6+1=	60+1=	6+1=
1+11=	3+0=	30+1=	8+1=
1+8=	8+1=	90+1=	2+1=
1+4=	1+3=	40+0=	11+1=
1+1=	40+1=	1+15=	15+1=
8+1=	12+1=	1+80=	1+16=
15+1=	0+5=	1+40=	7+1=
1+9=	1+11=	1+50=	4+1=
8+1=	1+6=	1+18=	9+1=
1+9=	1+4=	1+8=	3+1=
1+1=	12+0=	1+50=	6+1=

Lesson 31*   Minus 0s and 1s

Name: _______________________

11-1=	2-0=	3-1=	5-1=
4-1=	8-1=	5-1=	11-1=
9-0=	1-1=	10-1=	9-1=
3-1=	6-1=	8-1=	2-1=
7-1=	3-0=	6-1=	10-1=
4-1=	5-1=	3-1=	2-1=
11-1=	10-1=	5-0=	7-1=
9-1=	8-1=	4-1=	10-1=
2-1=	5-1=	11-1=	1-0=
6-0=	11-1=	10-1=	5-1=
3-1=	1-1=	5-1=	9-1=

Facts: −1, −0: b

Name: _______________________________

5-1=	7-1=	11-0=	4-1=
10-1=	4-1=	2-1=	7-1=
8-0=	9-1=	4-1=	10-1=
11-1=	10-1=	2-1=	3-1=
5-1=	6-0=	11-1=	1-1=
10-1=	3-1=	9-1=	5-1=
4-1=	4-0=	7-1=	9-0=
2-1=	7-1=	2-1=	11-1=
3-1=	11-1=	5-0=	1-1=
5-1=	9-1=	2-1=	8-1=
11-0=	10-1=	6-1=	4-1=

Lesson 33*   Plus/Minus 0s and 1s

Name: _______________________

5+0=	10-1=	16-1=	18+0=
7+1=	2+1=	14-0=	9-1=
6-1=	17+1=	15-1=	18+0=
3+1=	13-0=	17+1=	16-1=
10+0=	9-1=	12-0=	13-1=
2+0=	15-0=	7+1=	3-0=
1+1=	13-1=	1+0=	12+1=
5+1=	10-1=	6-1=	8+0=
3+1=	5-0=	3-1=	9-1=
9+0=	8+0=	4-1=	18-1=
1+1=	9-1=	2+1=	14+1=

Lesson 34   Plus/Minus 0s and 1s

Name: ___________________________

5+1=	10+1=	16-1=	18-1=
7-1=	2-0=	14+1=	9+1=
6-1=	17-1=	15+1=	18-1=
3+1=	14+1=	17-1=	16+1=
10+1=	9+1=	15+1=	13-0=
2-1=	15+1=	7-1=	3+1=
1+1=	13+1=	1-1=	12-1=
5+1=	10+1=	6-1=	8-1=
3-1=	5+1=	3+1=	9-1=
9+0=	8-1=	4+1=	14+1=
1+1=	9+1=	2-1=	14-1=

Lesson 35   Plus/Minus 0s and 1s   Facts: + − 0,1 mix: a

Name: _______________________

5+0=	10-1=	16-1=	18+0=
7+1=	2+1=	14-0=	9-1=
6-1=	17+1=	15-1=	18+0=
3+1=	13-0=	17+1=	16-1=
10+0=	9-1=	12-0=	13-1=
2+0=	15-0=	7+1=	3-0=
1+1=	13-1=	1+0=	12+1=
5+1=	10-1=	6-1=	8+0=
3+1=	5-0=	3-1=	9-1=
9+0=	8+0=	4-1=	18-1=
1+1=	9-1=	2+1=	14+1=

Name: _______________________________

5-1=	7-1=	11-0=	4-1=
10-1=	4-1=	2-1=	7-1=
8-0=	9-1=	4-1=	10-1=
11-1=	10-1=	2-1=	3-1=
5-1=	6-0=	11-1=	1-1=
10-1=	3-1=	9-1=	5-1=
4-1=	4-0=	7-1=	9-0=
2-1=	7-1=	2-1=	11-1=
3-1=	11-1=	5-0=	1-1=
5-1=	9-1=	2-1=	8-1=
11-0=	10-1=	6-1=	4-1=

Lesson 37   Plus 0s and 1s

Name: _______________________

1+5=	6+1=	60+1=	6+1=
1+11=	3+0=	30+1=	8+1=
1+8=	8+1=	90+1=	2+1=
1+4=	1+3=	40+0=	11+1=
1+1=	40+1=	1+15=	15+1=
8+1=	12+1=	1+80=	1+16=
15+1=	0+5=	1+40=	7+1=
1+9=	1+11=	1+50=	4+1=
8+1=	1+6=	1+18=	9+1=
1+9=	1+4=	1+8=	3+1=
1+1=	12+0=	1+50=	6+1=

Facts: −1, −0: a

Name: _______________________

11-1=	2-0=	3-1=	5-1=
4-1=	8-1=	5-1=	11-1=
9-0=	1-1=	10-1=	9-1=
3-1=	6-1=	8-1=	2-1=
7-1=	3-0=	6-1=	10-1=
4-1=	5-1=	3-1=	2-1=
11-1=	10-1=	5-0=	7-1=
9-1=	8-1=	4-1=	10-1=
2-1=	5-1=	11-1=	1-0=
6-0=	11-1=	10-1=	5-1=
3-1=	1-1=	5-1=	9-1=

Lesson 39 **T**   Plus/Minus 0s and 1s   Facts: + − 0, 1 mix: b

Name: _______________________

5+1=	10+1=	16-1=	18-1=
7-1=	2-0=	14+1=	9+1=
6-1=	17-1=	15+1=	18-1=
3+1=	14+1=	17-1=	16+1=
10+1=	9+1=	15+1=	13-0=
2-1=	15+1=	7-1=	3+1=
1+1=	13+1=	1-1=	12-1=
5+1=	10+1=	6-1=	8-1=
3-1=	5+1=	3+1=	9-1=
9+0=	8-1=	4+1=	14+1=
1+1=	9+1=	2-1=	14-1=

Lesson 40   Plus 0s and 1s

Facts: +1, 1+, 0 mix: a

Name: _______________________

7+1=	20+1=	1+4=	1+5=
4+1=	50+1=	0+12=	1+11=
9+1=	60+1=	1+6=	1+8=
3+1=	30+1=	1+9=	1+4=
6+0=	90+1=	1+3=	1+1=
8+1=	40+1=	1+7=	1+16=
2+1=	3+0=	1+4=	1+5=
11+1=	7+1=	1+6=	1+11=
1+1=	4+1=	0+10=	1+9=
1+16=	9+1=	50+1=	1+4=
30+1=	1+3=	1+16=	1+0=

Name: _______________________

2+3=	2+4=	2+1=	2+5=
2+2=	2+5=	2+3=	2+1=
2+0=	2+3=	2+0=	2+2=
2+5=	2+2=	2+1=	2+0=
2+2=	2+5=	2+2=	2+5=
2+4=	2+3=	2+4=	2+2=
2+5=	2+1=	2+0=	2+5=
2+3=	2+4=	2+2=	2+0=
2+2=	2+3=	2+5=	2+1=
2+1=	2+0=	2+3=	2+4=
2+5=	2+1=	2+2=	2+0=

Lesson 42

Plus 2s

Facts: 2+ (1–5): b

Name: ___________________________

2+1=	2+0=	2+2=	2+4=
2+3=	2+5=	2+1=	2+5=
2+4=	2+1=	2+2=	2+3=
2+2=	2+0=	2+4=	2+2=
2+1=	2+3=	2+5=	2+4=
2+5=	2+2=	2+1=	2+0=
2+3=	2+0=	2+2=	2+3=
2+2=	2+5=	2+4=	2+5=
2+3=	2+4=	2+0=	2+1=
2+5=	2+3=	2+1=	2+2=
2+4=	2+5=	2+3=	2+0=

Lesson 43*   Plus 2s

Name: ___________________

2+9=	2+8=	2+4=	2+10=
2+3=	2+4=	2+7=	2+5=
2+2=	2+6=	2+8=	2+1=
2+7=	2+8=	2+9=	2+4=
2+0=	2+10=	2+5=	2+2=
2+10=	2+2=	2+3=	2+6=
2+3=	2+7=	2+1=	2+2=
2+4=	2+9=	2+0=	2+3=
2+2=	2+3=	2+10=	2+8=
2+7=	2+10=	2+3=	2+2=
2+5=	2+0=	2+6=	2+9=

Lesson 44   Plus 2s

☐

Name: _______________________________

2+8=	2+5=	2+4=	2+10=
2+1=	2+4=	2+0=	2+4=
2+5=	2+3=	2+7=	2+8=
2+4=	2+5=	2+9=	2+6=
2+6=	2+0=	2+10=	2+7=
2+3=	2+2=	2+3=	2+5=
2+8=	2+3=	2+9=	2+7=
2+6=	2+1=	2+3=	2+2=
2+10=	2+4=	2+7=	2+6=
2+2=	2+9=	2+5=	2+7=
2+4=	2+0=	2+9=	2+2=

Lesson 45   Plus 2s

Name: _______________________

4+2=	0+2=	6+2=	5+2=
8+2=	2+2=	4+2=	8+2=
9+2=	5+2=	7+2=	6+2=
4+2=	8+2=	3+2=	2+2=
9+2=	1+2=	4+2=	6+2=
10+2=	3+2=	0+2=	5+2=
8+2=	4+2=	9+2=	7+2=
2+2=	3+2=	5+2=	10+2=
1+2=	7+2=	6+2=	4+2=
2+2=	4+2=	3+2=	6+2=
6+2=	3+2=	1+2=	10+2=

Lesson 46   Plus 2s

Name: _______________________________

2+2=	9+2=	7+2=	6+2=
5+2=	6+2=	4+2=	8+2=
2+2=	3+2=	10+2=	2+2=
9+2=	8+2=	7+2=	5+2=
4+2=	1+2=	8+2=	6+2=
5+2=	10+2=	6+2=	7+2=
8+2=	2+2=	5+2=	4+2=
5+2=	6+2=	8+2=	9+2=
6+2=	1+2=	0+2=	4+2=
8+2=	6+2=	10+2=	3+2=
5+2=	0+2=	2+2=	7+2=

Lesson 47*   Plus 2s

Name: _______________________

2+6=	2+3=	2+4=	2+5=
2+8=	2+7=	2+20=	2+8=
2+4=	2+9=	2+7=	2+3=
1+2=	4+2=	2+6=	2+1=
2+2=	20+2=	7+2=	5+2=
2+5=	60+2=	4+2=	2+4=
2+6=	30+2=	9+2=	2+8=
60+2=	90+2=	3+2=	2+9=
2+7=	9+2=	2+6=	90+2=
2+20=	1+2=	2+10=	2+8=
2+5=	2+9=	50+2=	2+5=

Lesson 48   Plus 2s

Name: _______________________

2+1=	2+3=	30+2=	2+4=
2+8=	2+7=	1+2=	2+6=
2+4=	2+10=	40+2=	2+2=
2+40=	2+80=	8+2=	20+2=
2+10=	2+3=	2+40=	2+5=
2+4=	2+40=	2+50=	2+1=
8+2=	2+2=	2+8=	2+9=
2+2=	2+5=	2+3=	2+90=
2+6=	2+2=	2+80=	1+2=
2+7=	2+9=	2+5=	5+2=
2+8=	2+5=	2+2=	2+6=

Lesson 49   Plus 2s

Name: _______________________

6+2=	3+2=	2+4=	2+5=
8+2=	7+2=	2+20=	2+2=
2+4=	40+2=	50+2=	2+3=
1+2=	4+2=	2+6=	2+1=
2+2=	20+2=	7+2=	2+2=
2+5=	60+2=	4+2=	2+4=
2+6=	30+2=	9+2=	2+8=
60+2=	90+2=	3+2=	2+9=
2+7=	9+2=	2+6=	90+2=
2+20=	1+2=	2+20=	2+8=
2+5=	2+9=	50+2=	2+5=

Lesson 50 **T**   Plus 2s

Name: ______________________________

2+1=	2+3=	30+2=	2+4=
2+8=	2+7=	1+2=	2+6=
2+4=	2+10=	40+2=	2+2=
2+40=	2+80=	9+2=	20+2=
2+10=	2+3=	2+40=	2+5=
2+4=	2+40=	2+50=	2+1=
8+2=	2+2=	2+8=	2+9=
2+2=	5+2=	3+2=	7+2=
6+2=	2+2=	4+2=	1+2=
2+7=	2+9=	2+5=	5+2=
2+8=	2+5=	2+2=	2+6=

Name: ___________________

11-2=	2-2=	3-2=	5-2=
4-2=	8-2=	5-2=	12-2=
7-2=	12-2=	10-2=	9-2=
3-2=	6-2=	8-2=	2-2=
7-2=	3-2=	6-2=	11-2=
4-2=	5-2=	3-2=	2-2=
11-2=	10-2=	5-2=	7-2=
9-2=	8-2=	4-2=	10-2=
2-2=	5-2=	11-2=	12-2=
6-2=	11-2=	10-2=	5-2=
3-2=	12-2=	5-2=	9-2=

Name: _______________________

5-2=	7-2=	12-2=	4-2=
10-2=	4-2=	2-2=	7-2=
8-2=	9-2=	4-2=	10-2=
11-2=	10-2=	2-2=	3-2=
5-2=	6-2=	11-2=	12-2=
10-2=	3-2=	9-2=	5-2=
4-2=	6-2=	7-2=	9-2=
2-2=	7-2=	2-2=	11-2=
3-2=	11-2=	5-2=	12-2=
5-2=	9-2=	2-2=	8-2=
12-2=	10-2=	6-2=	4-2=

Name: _______________________

3+2=	11+1=	5+1=	2+2=
5+2=	4+0=	12+2=	8+1=
10+2=	9+2=	9+1=	12+2=
8+1=	3+2=	2+1=	3+2=
6+2=	7+1=	11+2=	6+2=
3+2=	4+2=	2+0=	5+2=
5+0=	11+1=	7+2=	10+1=
4+2=	9+0=	10+2=	8+0=
11+2=	2+2=	12+2=	5+2=
10+0=	6+2=	5+1=	11+2=
8+1=	9+0=	3+2=	6+1=

Name: ___________________________

5+2=	3+1=	9+1=	1+2=
2+2=	5+2=	4+0=	7+1=
2+1=	10+2=	7+2=	4+2=
4+1=	8+2=	10+1=	9+2=
2+2=	11+2=	3+2=	10+0=
11+0=	5+1=	12+2=	6+2=
9+2=	10+0=	5+2=	3+2=
7+2=	4+2=	9+2=	4+2=
4+2=	2+2=	11+1=	7+0=
9+2=	9+1=	9+0=	6+2=
7+2=	10+1=	4+0=	3+2=

Lesson 55   Minus Mix: 0s, 1s, 2s   Facts: −2, −1, −0 mix: a

Name: _______________________

11-1=	2-2=	3-2=	5-1=
4-0=	8-1=	5-2=	12-2=
9-2=	12-2=	10-2=	9-1=
3-2=	6-2=	8-1=	2-1=
7-1=	3-2=	6-2=	11-2=
4-2=	5-2=	3-2=	2-0=
11-1=	10-1=	5-0=	7-2=
9-0=	8-0=	4-2=	10-2=
2-2=	5-2=	11-2=	12-2=
6-2=	11-2=	10-0=	5-1=
3-1=	12-2=	5-2=	9-1=

Lesson 56   Minus Mix: 0s, 1s, 2s   Facts: −2, −1, −0 mix: b

5-2=	7-1=	12-2=	4-0=
10-2=	4-1=	2-1=	7-2=
8-2=	9-2=	4-1=	10-1=
11-2=	10-0=	2-2=	3-2=
5-1=	6-2=	11-0=	12-2=
10-0=	3-2=	9-2=	5-2=
4-1=	4-2=	7-2=	9-2=
2-2=	7-0=	2-2=	11-1=
3-0=	11-2=	5-1=	12-2=
5-1=	9-2=	2-2=	8-1=
12-2=	10-1=	6-2=	4-1=

Lesson 57 T Minus Mix: 0s, 1s, 2s Facts: −2, −1, −0 mix: a

Name: _______________________________

11-1=	2-2=	3-2=	5-1=
4-0=	8-1=	5-2=	12-2=
9-2=	12-2=	10-2=	9-1=
3-2=	6-2=	8-1=	2-1=
7-1=	3-2=	6-2=	11-2=
4-2=	5-2=	3-2=	2-0=
11-1=	10-1=	5-0=	7-2=
9-0=	8-0=	4-2=	10-2=
2-2=	5-2=	11-2=	12-2=
6-2=	11-2=	10-0=	5-1=
3-1=	12-2=	5-2=	9-1=

Facts: +2, +1, +0 mix: b

Name: _______________________

5+2=	3+1=	9+1=	1+2=
2+2=	5+2=	4+0=	7+1=
2+1=	10+2=	7+2=	4+2=
4+1=	8+2=	10+1=	9+2=
2+2=	11+2=	3+2=	10+0=
11+0=	5+1=	12+2=	6+2=
9+2=	10+0=	5+2=	3+2=
7+2=	4+2=	9+2=	4+2=
4+2=	2+2=	11+1=	7+0=
9+2=	9+1=	9+0=	6+2=
7+2=	10+1=	4+0=	3+2=

Name: _______________

3+2=	11+1=	5+1=	2+2=
5+2=	4+0=	12+2=	8+1=
10+2=	9+2=	9+1=	12+2=
8+1=	3+2=	2+1=	3+2=
6+2=	7+1=	11+2=	6+2=
3+2=	4+2=	2+0=	5+2=
5+0=	11+1=	7+2=	10+1=
4+2=	9+0=	10+2=	8+0=
11+2=	2+2=	12+2=	5+2=
10+0=	6+2=	5+1=	11+2=
8+1=	9+0=	3+2=	6+1=

Lesson 60   Minus Mix: 0s, 1s, 2s

Facts: −2, −1, −0 mix: b

Name: _______________________

5-2=	7-1=	12-2=	4-0=
10-2=	4-1=	2-1=	7-2=
8-2=	9-2=	4-1=	10-1=
11-2=	10-0=	2-2=	3-2=
5-1=	6-2=	11-0=	12-2=
10-0=	3-2=	9-2=	5-2=
4-1=	4-2=	7-2=	9-2=
2-2=	7-0=	2-2=	11-1=
3-0=	11-2=	5-1=	12-2=
5-1=	9-2=	2-2=	8-1=
12-2=	10-1=	6-2=	4-1=

Lesson 61*   Plus 10s

Name: _______________________

9+10=   4+10=   6+10=   5+10=

3+10=   5+10=   3+10=   7+10=

1+10=   7+10=   10+10=   1+10=

10+10=   5+10=   2+10=   6+10=

4+10=   7+10=   8+10=   3+10=

5+10=   1+10=   9+10=   10+10=

7+10=   6+10=   5+10=   2+10=

6+10=   3+10=   1+10=   7+10=

3+10=   4+10=   2+10=   8+10=

10+10=   2+10=   5+10=   9+10=

2+10=   8+10=   7+10=   8+10=

Lesson 62   Plus 10s

Name: ___________________

10+5=	10+6=	10+4=	10+5=
10+9=	10+7=	10+10=	10+7=
10+8=	10+3=	10+6=	10+8=
10+4=	10+8=	10+9=	10+4=
10+7=	10+2=	10+3=	10+5=
10+6=	10+4=	10+7=	10+10=
10+9=	10+6=	10+10=	10+5=
10+6=	10+7=	10+5=	10+9=
10+6=	10+4=	10+8=	10+3=
10+4=	10+6=	10+4=	10+2=
10+7=	10+10=	10+5=	10+1=

Lesson 63   Plus 10s

Name: _______________________

7+10=	2+10=	10+4=	10+5=
4+10=	6+10=	10+2=	10+10=
9+10=	3+10=	5+10=	10+3=
3+10=	9+10=	10+6=	10+10=
6+10=	3+10=	10+9=	10+2=
8+10=	7+10=	10+3=	10+4=
2+10=	4+10=	10+7=	10+5=
10+10=	3+10=	10+4=	10+6=
0+10=	9+10=	10+7=	6+10=
10+6=	10+10=	10+3=	10+7=
1+10=	10+5=	8+10=	10+2=

Name: _______________________

10+5=	10+8=	5+10=	10+8=
10+10=	10+3=	3+10=	10+9=
10+8=	10+4=	10+2=	9+10=
10+3=	10+10=	4+10=	8+10=
10+7=	10+5=	8+10=	10+9=
10+5=	10+10=	10+2=	10+4=
10+0=	10+7=	10+5=	10+6=
8+10=	10+8=	10+6=	10+10=
10+5=	4+10=	10+3=	6+10=
10+6=	10+2=	10+8=	10+5=
10+7=	6+10=	3+10=	10+10=

Lesson 65   Plus Mix: 0s, 1s, 2s, 10s

Name: _______________________

10 + 2	3 + 1	5 + 2	2 + 0	6 + 1
4 + 1	2 + 9	2 + 6	1 +10	5 + 0
7 + 2	2 + 6	5 + 0	1 + 3	9 + 1
7 +10	0 + 8	6 + 1	2 + 1	2 + 3
2 + 4	7 + 2	10 + 2	3 + 0	2 + 7
3 + 2	10 + 3	9 + 0	1 + 5	8 + 10
5 +10	6 + 2	9 + 1	1 + 4	3 + 0
4 + 1	7 + 1	9 + 2	0 + 5	10 + 2
9 +10	1 + 1	0 + 0	1 + 3	9 + 1

Lesson 66   Plus Mix: 0s, 1s, 2s, 10s

Name: _______________________

10+7=	2+4=	0+3=	9+0=
5+10=	3+1=	1+6=	6+10=
10+3=	6+2=	5+2=	2+8=
1+0=	5+1=	7+2=	2+4=
6+1=	1+3=	0+9=	9+0=
7+10=	10+3=	2+4=	7+2=
4+1=	0+7=	5+2=	10+3=
10+4=	2+3=	6+2=	4+10=
8+10=	0+4=	1+10=	5+0=
4+1=	1+8=	7+0=	1+9=
10+6=	4+10=	1+1=	7+2=

Lesson 67   Plus Mix: 0s, 1s, 2s, 10s

Facts: +0, 1, 2, 10 mix: c

Name: _______________________

9 + 2	2 + 8	7 + 1	10 + 4	9 + 2
1 + 6	5 + 0	2 +10	2 + 6	6 + 2
0 + 7	4 + 1	0 + 2	7 +10	1 + 4
6 + 1	4 +10	10 + 1	2 + 6	0 + 6
5 + 1	0 + 8	5 + 2	10 + 8	1 + 4
7 +10	1 + 6	9 + 1	0 + 5	2 +10
10 + 0	6 + 1	7 +10	2 + 3	10 + 3
1 + 7	6 + 2	2 + 0	5 + 1	1 + 5
0 + 4	4 + 1	10 + 6	5 +10	4 + 0

Lesson 68   Plus Mix: 0s, 1s, 2s, 10s   Facts: +0, 1, 2 mix

Name: _________________________

5+1=	11+1=	3+2=	2+2=
12+2=	4+0=	5+2=	8+1=
9+1=	9+2=	10+2=	12+2=
2+1=	3+2=	8+1=	6+2=
1+2=	7+1=	6+2=	3+2=
2+0=	4+2=	3+2=	5+2=
7+2=	11+1=	5+0=	10+1=
10+2=	9+0=	4+2=	8+0=
12+2=	2+2=	11+2=	5+2=
5+1=	6+2=	10+0=	11+2=
9+1=	3+1=	5+2=	12+2=

Lesson 69 T   Plus Mix: 0s, 1s, 2s, 10s Facts: +0, 1, 2 mix

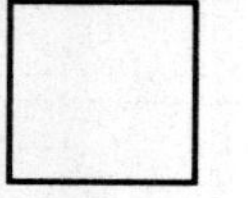

9 + 1	3 + 1	5 + 2	12 + 2	11 + 2
4 + 0	5 + 2	2 + 2	7 + 1	9 + 1
7 + 2	10 + 2	2 + 1	4 + 1	3 + 1
10 + 1	8 + 2	4 + 1	9 + 2	5 + 2
3 + 2	11 + 2	2 + 2	10 + 0	12 + 2
1 + 2	5 + 1	11 + 0	6 + 2	11 + 0
5 + 2	10 + 0	9 + 2	3 + 2	7 + 1
9 + 2	4 + 2	7 + 2	4 + 2	6 + 2
11 + 1	2 + 2	10 + 2	7 + 0	8 + 1

Lesson 70   Minus Mix: 0s, 1s, 2s

Facts: −0, 1, 2 mix

Name: _______________________

11-1=	2-2=	3-2=	5-1=
4-0=	8-1=	5-2=	12-2=
9-2=	12-2=	10-2=	9-1=
3-2=	6-2=	8-1=	2-1=
7-1=	3-2=	6-2=	11-2=
4-2=	5-2=	3-2=	2-0=
11-1=	10-1=	5-0=	7-0=
9-0=	8-0=	4-2=	10-2=
2-2=	5-2=	11-2=	12-2=
6-2=	11-2=	10-0=	5-1=
3-1=	12-2=	5-2=	9-1=

Lesson 71　Minus Mix: 0s, 1s, 2s

Name: _______________________

$$\begin{array}{r} 5 \\ -\ 2 \\ \hline \end{array} \qquad \begin{array}{r} 7 \\ -\ 1 \\ \hline \end{array} \qquad \begin{array}{r} 12 \\ -\ 2 \\ \hline \end{array} \qquad \begin{array}{r} 4 \\ -\ 0 \\ \hline \end{array} \qquad \begin{array}{r} 4 \\ -\ 2 \\ \hline \end{array}$$

$$\begin{array}{r} 10 \\ -\ 2 \\ \hline \end{array} \qquad \begin{array}{r} 4 \\ -\ 1 \\ \hline \end{array} \qquad \begin{array}{r} 2 \\ -\ 1 \\ \hline \end{array} \qquad \begin{array}{r} 7 \\ -\ 2 \\ \hline \end{array} \qquad \begin{array}{r} 11 \\ -\ 1 \\ \hline \end{array}$$

$$\begin{array}{r} 8 \\ -\ 2 \\ \hline \end{array} \qquad \begin{array}{r} 9 \\ -\ 2 \\ \hline \end{array} \qquad \begin{array}{r} 4 \\ -\ 1 \\ \hline \end{array} \qquad \begin{array}{r} 10 \\ -\ 1 \\ \hline \end{array} \qquad \begin{array}{r} 8 \\ -\ 0 \\ \hline \end{array}$$

$$\begin{array}{r} 11 \\ -\ 2 \\ \hline \end{array} \qquad \begin{array}{r} 10 \\ -\ 0 \\ \hline \end{array} \qquad \begin{array}{r} 2 \\ -\ 2 \\ \hline \end{array} \qquad \begin{array}{r} 3 \\ -\ 2 \\ \hline \end{array} \qquad \begin{array}{r} 3 \\ -\ 0 \\ \hline \end{array}$$

$$\begin{array}{r} 5 \\ -\ 1 \\ \hline \end{array} \qquad \begin{array}{r} 6 \\ -\ 2 \\ \hline \end{array} \qquad \begin{array}{r} 11 \\ -\ 0 \\ \hline \end{array} \qquad \begin{array}{r} 12 \\ -\ 2 \\ \hline \end{array} \qquad \begin{array}{r} 10 \\ -\ 1 \\ \hline \end{array}$$

$$\begin{array}{r} 10 \\ -\ 0 \\ \hline \end{array} \qquad \begin{array}{r} 3 \\ -\ 2 \\ \hline \end{array} \qquad \begin{array}{r} 9 \\ -\ 2 \\ \hline \end{array} \qquad \begin{array}{r} 5 \\ -\ 2 \\ \hline \end{array} \qquad \begin{array}{r} 6 \\ -\ 1 \\ \hline \end{array}$$

$$\begin{array}{r} 4 \\ -\ 2 \\ \hline \end{array} \qquad \begin{array}{r} 4 \\ -\ 1 \\ \hline \end{array} \qquad \begin{array}{r} 7 \\ -\ 2 \\ \hline \end{array} \qquad \begin{array}{r} 9 \\ -\ 2 \\ \hline \end{array} \qquad \begin{array}{r} 5 \\ -\ 2 \\ \hline \end{array}$$

$$\begin{array}{r} 2 \\ -\ 2 \\ \hline \end{array} \qquad \begin{array}{r} 7 \\ -\ 0 \\ \hline \end{array} \qquad \begin{array}{r} 2 \\ -\ 2 \\ \hline \end{array} \qquad \begin{array}{r} 11 \\ -\ 1 \\ \hline \end{array} \qquad \begin{array}{r} 4 \\ -\ 1 \\ \hline \end{array}$$

$$\begin{array}{r} 3 \\ -\ 0 \\ \hline \end{array} \qquad \begin{array}{r} 11 \\ -\ 2 \\ \hline \end{array} \qquad \begin{array}{r} 5 \\ -\ 1 \\ \hline \end{array} \qquad \begin{array}{r} 12 \\ -\ 2 \\ \hline \end{array} \qquad \begin{array}{r} 9 \\ -\ 2 \\ \hline \end{array}$$

Lesson 72   Plus 10s

Name: _______________________

7+10=	2+10=	10+4=	10+5=
4+10=	6+10=	10+2=	10+10=
9+10=	3+10=	5+10=	10+3=
3+10=	9+10=	10+6=	10+10=
6+10=	3+10=	10+9=	10+2=
8+10=	7+10=	10+3=	10+4=
2+10=	4+10=	10+7=	10+5=
10+10=	3+10=	10+4=	10+6=
0+10=	9+10=	10+7=	6+10=
10+6=	10+10=	10+3=	10+7=
1+10=	10+5=	8+10=	10+2=

Lesson 73   Plus Mix: 0s, 1s, 2s, 10s

Name: ___________________

9 + 2	2 + 8	7 + 1	10 + 4	9 + 2
1 + 6	5 + 0	2 +10	2 + 6	6 + 2
0 + 7	4 + 1	0 + 2	7 +10	1 + 4
6 +1	4 +10	10 + 1	2 + 6	0 + 6
5 + 1	0 + 8	5 + 2	10 + 8	1 + 4
7 +10	1 + 6	9 + 1	0 + 5	2 +10
10 + 0	6 + 1	7 +10	2 + 3	10 + 3
1 + 7	6 + 2	2 + 0	5 + 1	1 + 5
0 + 4	4 + 1	10 + 6	5 +10	4 + 0

Lesson 74   Minus Mix: 0s, 1s, 2s

Facts: −0, 1, 2, mix

Name: _______________________

5 − 2	7 − 1	12 − 2	4 − 0	4 − 2
10 − 2	4 − 1	2 − 1	7 − 2	11 − 1
8 − 2	9 − 2	4 − 1	10 − 1	8 − 0
11 − 2	10 − 0	2 − 2	3 − 2	3 − 0
5 − 1	6 − 2	11 − 0	12 − 2	10 − 1
10 − 0	3 − 2	9 − 2	5 − 2	6 − 1
4 − 2	4 − 1	7 − 2	9 − 2	5 − 2
2 − 2	7 − 0	2 − 2	11 − 1	4 − 1
3 − 0	11 − 2	5 − 1	12 − 2	9 − 2

Lesson 75*   Plus 9s

Facts: +9: a

Name: _______________________________

1+9=	2+9=	9+9=	8+9=
4+9=	10+9=	3+9=	5+9=
7+9=	6+9=	8+9=	4+9=
3+9=	5+9=	10+9=	7+9=
6+9=	9+9=	1+9=	2+9=
8+9=	4+9=	2+9=	1+9=
2+9=	7+9=	8+9=	3+9=
10+9=	3+9=	6+9=	5+9=
5+9=	8+9=	3+9=	6+9=
1+9=	10+9=	2+9=	4+9=
8+9=	4+9=	7+9=	2+9=

Lesson 76*   Plus 9s

Name: ___________________________

4+9=	2+9=	9+9 =	9+4=
1+9=	10+9=	9+3=	9+90=
5+9=	6+9=	9+6=	9+8=
3+9=	5+9=	9+5=	3+9=
6+9=	9+9=	9+2=	9+1=
8+9=	4+9=	9+7=	9+10=
2+9=	7+9=	9+4=	9+5=
10+9=	3+9=	9+6=	9+3=
5+9=	8+9=	9+7=	9+8=
1+9=	10+9=	3+9=	4+9=
2+9=	9+3=	6+9=	9+1=

Lesson 77   Plus 9s

Name: _______________________________

9+4=	9+3=	6+9=	9+9=
9+8=	2+9=	9+6=	9+10=
9+9=	5+9=	9+9=	9+3=
5+9=	7+9=	9+4=	9+2=
9+9=	30+9=	9+7=	9+6=
6+9=	3+9=	2+9=	9+4=
8+9=	10+9=	9+6=	9+1=
2+9=	5+9=	9+10=	3+9=
1+9=	7+9=	10+9=	9+3=
3+9=	4+9=	3+9=	6+9=
9+8=	2+9=	9+6=	9+10=

Lesson 78*   Plus 9s

Name: _______________________

9+6=	5+9=	9+9=	0+9=
9+2=	6+9=	4+9=	8+9=
9+1=	3+9=	10+9=	2+9=
9+8=	8+9=	9+3=	9+4=
9+9=	1+9=	9+7=	9+6=
9+10=	4+9=	9+3=	7+9=
8+9=	2+9=	9+10=	4+9=
3+9=	9+9=	9+8=	5+9=
2+9=	9+1=	9+80=	9+4=
8+9=	9+6=	10+9=	3+9=
9+0=	4+9=	3+9=	9+7=

Lesson 79   Plus 9s

Name: ________________________

4+9=	2+9=	9+9=	9+4=
1+9=	10+9=	9+3=	9+90=
5+9=	6+9=	9+6=	9+8=
3+9=	5+9=	9+5=	3+9=
6+9=	9+9=	9+2=	9+1=
8+9=	4+9=	9+7=	9+10=
2+9=	7+9=	9+4=	9+5=
10+9=	3+9=	9+6=	9+3=
5+9=	8+9=	9+7=	9+8=
1+9=	10+9=	3+9=	4+9=
2+9=	9+3=	6+9=	9+1=

Facts: 9+ +9: c

Name: ___________________________

9+6=	5+9=	9+9=	0+9=
9+2=	6+9=	4+9=	8+9=
9+1=	3+9=	10+9=	2+9=
9+8=	8+9=	9+3=	9+4=
9+9=	1+9=	9+7=	9+6=
9+10=	4+9=	9+3=	7+9=
8+9=	2+9=	9+10=	4+9=
3+9=	9+9=	9+8=	5+9=
2+9=	9+1=	9+80=	9+4=
8+9=	9+6=	10+9=	3+9=
9+0=	4+9=	3+9=	9+7=

Lesson 81*   Minus 10s

Name: _______________________________

15-10=	10-10=	16-10=	18-10=
17-10=	12-10=	14-10=	19-10=
16-10=	17-10=	15-10=	18-10=
13-10=	14-10=	17-10=	16-10=
10-10=	19-10=	15-10=	13-10=
12-10=	15-10=	17-10=	14-10=
11-10=	13-10=	10-10=	12-10=
15-10=	10-10=	16-10=	18-10=
13-10=	15-10=	17-10=	19-10=
19-10=	18-10=	14-10=	17-10=
16-10=	19-10=	12-10=	14-10=

Lesson 82*   Minus Mix: 0s, 1s, 2s, 10s

Facts: -0, 1, 2, 10 mix

Name: _______________________________

5-0=	10-2=	16-1=	18-1=
7-1=	14-10=	17-10=	19-10=
16-10=	2-1=	14-2=	8-1=
3-1=	7-1=	15-10=	16-10=
10-2=	10-10=	17-1=	15-10=
2-1=	17-0=	18-0=	13-0=
11-10=	14-2=	14-10=	18-10=
5-0=	9-2=	10-10=	12-1=
3-1=	15-2=	18-10=	8-1=
19-10=	13-10=	7-1=	3-0=
12-2=	18-10=	10-1=	14-2=

Connecting Math Concepts Math Facts Worksheets, Level B

Lesson 83   Minus Mix: 0s, 1s, 2s, 10s

Facts: -0, 1, 2, 10 mix

Name: _______________________

19 − 10	13 − 10	7 − 1	3 − 0	15 − 10
12 − 2	18 − 10	10 − 1	14 − 2	10 − 0
15 − 10	14 − 0	16 − 10	14 − 1	8 − 2
10 − 1	5 − 2	13 − 10	17 − 10	7 − 0
7 − 1	8 − 1	14 − 10	19 − 10	6 − 2
17 − 10	19 − 10	2 − 1	10 − 10	18 − 1
18 − 10	15 − 10	3 − 1	11 − 1	5 − 2
14 − 10	4 − 0	10 − 2	17 − 1	4 − 1
10 − 10	11 − 1	15 − 0	15 − 1	13 − 10

Lesson 84 T   Minus Mix: 0s, 1s, 2s, 10s

Facts: -0, 1, 2, 10 mix

Name: _______________________

5-0=	10-2=	16-1=	18-1=
7-1=	14-10=	17-10=	19-10=
16-10=	2-1=	14-2=	8-1=
3-1=	7-1=	15-10=	16-10=
10-2=	10-10=	17-1=	15-10=
2-1=	17-0=	18-0=	13-0=
11-10=	14-2=	14-10=	18-10=
5-0=	9-2=	10-10=	12-1=
3-1=	15-2=	18-10=	8-1=
19-10=	13-10=	7-1=	3-0=
12-2=	18-10=	10-1=	14-2=

Lesson 85*   Plus 3s

Facts: 3+ (0–5): a

Name: ___________________

3+5=	3+0=	3+2=	3+4=
3+3=	3+1=	3+4=	3+5=
3+4=	3+0=	3+1=	3+3=
3+2=	3+3=	3+0=	3+4=
3+0=	3+1=	3+5=	3+2=
3+1=	3+2=	3+4=	3+0=
3+4=	3+3=	3+1=	3+4=
3+3=	3+5=	3+4=	3+1=
3+2=	3+4=	3+0=	3+3=
3+5=	3+3=	3+4=	3+1=
3+4=	3+5=	3+2=	3+3=

Name: _______________________

3+6=	3+1=	3+5=	3+7=
3+10=	3+0=	3+2=	3+4=
3+3=	3+10=	3+7=	3+5=
3+4=	3+7=	3+8=	3+3=
3+2=	3+3=	3+9=	3+8=
3+7=	3+10=	3+5=	3+2=
3+1=	3+2=	3+7=	3+0=
3+3=	3+6=	3+8=	3+7=
3+8=	3+5=	3+4=	3+10=
3+9=	3+4=	3+0=	3+2=
3+7=	3+2=	3+1=	3+4=

Lesson 87 **T**   **Plus 3s**

Name: _______________________

3+5=	3+3=	3+7=	3+8=
3+4=	3+5=	3+9=	3+5=
3+6=	3+2=	3+10=	3+7=
3+3=	3+8=	3+3=	3+5=
3+8=	3+4=	3+9=	3+7=
3+6=	3+6=	3+3=	3+2=
3+10=	3+8=	3+7=	3+6=
3+2=	3+10=	3+5=	3+0=
3+4=	3+3=	3+6=	3+2=
3+5=	3+10=	3+7=	3+4=
3+6=	3+9=	3+3=	3+8=

Lesson 88* Plus 3s

Name: ___________________________

8+3=	2+3=	4+3=	3+3=
4+3=	5+3=	0+3=	8+3=
9+3=	6+3=	10+3=	7+3=
5+3=	3+3=	7+3=	2+3=
6+3=	9+3=	0+3=	4+3=
8+3=	10+3=	4+3=	20+3=
2+3=	4+3=	5+3=	10+3=
7+3=	3+3=	8+3=	7+3=
5+3=	8+3=	6+3=	10+3=
1+3=	9+3=	4+3=	6+3=
8+3=	5+3=	7+3=	10+3=

Lesson 89   Plus Mix: 1s, 2s, 3s

Name: _______________________

9+3=	11+3=	5+3=	7+3=
5+2=	4+1=	12+2=	8+3=
10+2=	9+2=	11+3=	12+2=
8+3=	9+3=	2+3=	6+2=
6+2=	7+3=	11+2=	9+3=
9+2=	8+3=	2+1=	5+2=
5+1=	11+3=	7+2=	10+3=
8+3=	9+1=	10+2=	8+1=
11+2=	7+3=	12+2=	5+2=
10+1=	6+2=	5+3=	11+2=
5+2=	3+3=	9+3=	12+2=

Lesson 90   Plus 9s

Name: _______________________

9+6=	5+9=	9+9=	0+9=
9+2=	6+9=	4+9=	8+9=
9+1=	3+9=	10+9=	2+9=
9+8=	8+9=	9+3=	9+4=
9+9=	1+9=	9+7=	9+6=
9+10=	4+9=	9+3=	7+9=
8+9=	2+9=	9+10=	4+9=
3+9=	9+9=	9+8=	5+9=
2+9=	9+1=	9+80=	9+4=
8+9=	9+6=	10+9=	3+9=
9+0=	4+9=	3+9=	9+7=

Lesson 91*   Minus 9s

Name: _______________________________

9-9=	15-9=	19-9=	17-9=
12-9=	16-9=	13-9=	14-9=
18-9=	10-9=	9-9=	13-9=
11-9=	15-9=	18-9=	10-9=
15-9=	14-9=	17-9=	18-9=
13-9=	17-9=	14-9=	11-9=
19-9=	12-9=	16-9=	15-9=
13-9=	16-9=	10-9=	13-9=
17-9=	15-9=	18-9=	9-9=
14-9=	11-9=	16-9=	15-9=
13-9=	14-9=	11-9=	18-9=

Lesson 92*   Minus 9s

Name: _______________________

10-9=	16-9=	12-9=	15-9=
19-9=	14-9=	13-9=	16-9=
11-9=	13-9=	15-9=	12-9=
16-9=	15-9=	9-9=	17-9=
12-9=	13-9=	10-9=	9-9=
15-9=	11-9=	17-9=	12-9=
13-9=	18-9=	14-9=	16-9=
17-9=	19-9=	15-9=	13-9=
16-9=	13-9=	9-9=	18-9=
12-9=	16-9=	11-9=	13-9=
10-9=	13-9=	15-9=	19-9=

Name: ______________________

11-9=	15-9=	9-9=	13-9=
15-9=	14-9=	13-9=	16-9=
9-9=	16-9=	18-9=	11-9=
14-9=	10-9=	12-9=	13-9=
19-9=	18-9=	17-9=	14-9=
11-9=	17-9=	14-9=	18-9=
9-9=	14-9=	16-9=	10-9=
13-9=	16-9=	10-9=	12-9=
17-9=	15-9=	19-9=	17-9=
14-9=	11-9=	16-9=	9-9=
16-9=	14-9=	11-9=	15-9=

Lesson 94 T   Minus 9s

Name: _______________________

10-9=	16-9=	14-9=	18-9=
19-9=	14-9=	18-9=	10-9=
11-9=	13-9=	15-9=	14-9=
16-9=	15-9=	9-9=	12-9=
17-9=	18-9=	14-9=	16-9=
15-9=	11-9=	10-9=	9-9=
13-9=	9-9=	17-9=	11-9=
14-9=	19-9=	15-9=	16-9=
17-9=	13-9=	18-9=	15-9=
12-9=	16-9=	11-9=	19-9=
10-9=	13-9=	15-9=	9-9=

Lesson 95 Plus Mix: 1s, 2s, 3s

Facts: +1, +2, +3 mix: b

Name: _______________

7 + 3	5 + 2	14 + 1	7 + 3	11 + 2
2 + 3	10 + 3	7 + 2	4 + 3	4 + 1
9 + 3	8 + 2	10 + 3	9 + 2	6 + 3
7 + 3	11 + 2	9 + 3	10 + 1	10 + 3
11 + 1	5 + 3	7 + 3	6 + 2	5 + 2
9 + 2	10 + 1	5 + 2	9 + 3	3 + 3
7 + 2	11 + 3	9 + 2	8 + 3	7 + 1
7 + 3	10 + 2	11 + 3	7 + 1	8 + 1
2 + 2	3 + 1	10 + 1	9 + 2	5 + 3

Name: _______________________

$9+[\ ]=11$ $9+[\ ]=15$ $9+[\ ]=9$ $9+[\ ]=13$

$9+[\ ]=15$ $9+[\ ]=14$ $9+[\ ]=13$ $9+[\ ]=16$

$9+[\ ]=9$ $9+[\ ]=16$ $9+[\ ]=18$ $9+[\ ]=14$

$9+[\ ]=17$ $9+[\ ]=10$ $9+[\ ]=12$ $9+[\ ]=13$

$9+[\ ]=19$ $9+[\ ]=18$ $9+[\ ]=16$ $9+[\ ]=15$

$9+[\ ]=11$ $9+[\ ]=17$ $9+[\ ]=14$ $9+[\ ]=18$

$9+[\ ]=9$ $9+[\ ]=14$ $9+[\ ]=16$ $9+[\ ]=10$

$9+[\ ]=13$ $9+[\ ]=16$ $9+[\ ]=10$ $9+[\ ]=12$

$9+[\ ]=17$ $9+[\ ]=15$ $9+[\ ]=19$ $9+[\ ]=17$

$9+[\ ]=14$ $9+[\ ]=11$ $9+[\ ]=16$ $9+[\ ]=9$

$9+[\ ]=16$ $9+[\ ]=14$ $9+[\ ]=11$ $9+[\ ]=15$

Lesson 97*   9 Plus Algebra

Facts: alg 9+: b

Name: _________________________

9+[ ]=10   9+[ ]=16   9+[ ]=17   9+[ ]=18

9+[ ]=19   9+[ ]=14   9+[ ]=18   9+[ ]=10

9+[ ]=11   9+[ ]=13   9+[ ]=15   9+[ ]=14

9+[ ]=16   9+[ ]=15   9+[ ]=9   9+[ ]=19

9+[ ]=12   9+[ ]=13   9+[ ]=14   9+[ ]=16

9+[ ]=15   9+[ ]=12   9+[ ]=10   9+[ ]=9

9+[ ]=13   9+[ ]=9   9+[ ]=17   9+[ ]=11

9+[ ]=14   9+[ ]=19   9+[ ]=15   9+[ ]=16

9+[ ]=17   9+[ ]=13   9+[ ]=18   9+[ ]=15

9+[ ]=12   9+[ ]=16   9+[ ]=11   9+[ ]=19

9+[ ]=10   9+[ ]=13   9+[ ]=15   9+[ ]=9

Lesson 98   Plus 9s

Name: ___________________________

9+4 =	9+3=	6+9=	9+9=
9+8=	2+9=	9+6=	9+10=
9+9=	5+9=	9+9=	9+3=
5+9=	7+9=	9+4=	9+2=
9+9=	30+9=	9+7=	9+6=
6+9=	3+9=	2+9=	9+4=
8+9=	10+9=	9+6=	9+1=
2+9=	5+9=	9+10=	3+9=
1+9=	7+9=	10+9=	9+3=
3+9=	4+9=	3+9=	6+9=
9+8=	2+9=	9+6=	9+10=

Lesson 99   Plus 3s

Name: _______________________________

8+3=	2+3=	4+3=	3+3=
4+3=	5+3=	0+3=	8+3=
9+3=	6+3=	10+3=	7+3=
5+3=	3+3=	7+3=	2+3=
6+3=	9+3=	0+3=	4+3=
8+3=	10+3=	4+3=	20+3=
2+3=	4+3=	5+3=	10+3=
7+3=	3+3=	8+3=	7+3=
5+3=	8+3=	6+3=	10+3=
1+3=	9+3=	4+3=	6+3=
8+3=	5+3=	7+3=	10+3=

Facts: −9: b

Name: _______________________

10-9=	16-9=	12-9=	15-9=
19-9=	14-9=	13-9=	16-9=
11-9=	13-9=	15-9=	12-9=
16-9=	15-9=	9-9=	17-9=
12-9=	13-9=	10-9=	9-9=
15-9=	11-9=	17-9=	12-9=
13-9=	18-9=	14-9=	16-9=
17-9=	19-9=	15-9=	13-9=
16-9=	13-9=	9-9=	18-9=
12-9=	16-9=	11-9=	13-9=
10-9=	13-9=	15-9=	19-9=

Lesson 101   Minus 9s

Name: ___________________________

10-9=	16-9=	14-9=	18-9=
19-9=	14-9=	18-9=	10-9=
11-9=	13-9=	15-9=	14-9=
16-9=	15-9=	9-9=	12-9=
17-9=	18-9=	14-9=	16-9=
15-9=	11-9=	10-9=	9-9=
13-9=	9-9=	17-9=	11-9=
14-9=	19-9=	15-9=	16-9=
17-9=	13-9=	18-9=	15-9=
12-9=	16-9=	11-9=	19-9=
10-9=	13-9=	15-9=	9-9=

Lesson 102   9 Plus Algebra

Name: _______________________

9+[ ]=11	9+[ ]=15	9+[ ]=9	9+[ ]=13
9+[ ]=15	9+[ ]=14	9+[ ]=13	9+[ ]=16
9+[ ]=9	9+[ ]=16	9+[ ]=18	9+[ ]=14
9+[ ]=17	9+[ ]=10	9+[ ]=12	9+[ ]=13
9+[ ]=19	9+[ ]=18	9+[ ]=16	9+[ ]=15
9+[ ]=11	9+[ ]=17	9+[ ]=14	9+[ ]=18
9+[ ]=9	9+[ ]=14	9+[ ]=16	9+[ ]=10
9+[ ]=13	9+[ ]=16	9+[ ]=10	9+[ ]=12
9+[ ]=17	9+[ ]=15	9+[ ]=19	9+[ ]=17
9+[ ]=14	9+[ ]=11	9+[ ]=16	9+[ ]=9
9+[ ]=16	9+[ ]=14	9+[ ]=11	9+[ ]=15

Lesson 103   Plus 9s

Name: _______________________________

9+6=	5+9=	9+9=	0+9=
9+2=	6+9=	4+9=	8+9=
9+1=	3+9=	10+9=	2+9=
9+8=	8+9=	9+3=	9+4=
9+9=	1+9=	9+7=	9+6=
9+10=	4+9=	9+3=	7+9=
8+9=	2+9=	9+10=	4+9=
3+9=	9+9=	9+8=	5+9=
2+9=	9+1=	9+80=	9+4=
8+9=	9+6=	10+9=	3+9=
9+0=	4+9=	3+9=	9+7=

Lesson 104 9 Plus Algebra

Name: _______________________________

9+[ ]=10	9+[ ]=16	9+[ ]=17	9+[ ]=18
9+[ ]=19	9+[ ]=14	9+[ ]=18	9+[ ]=10
9+[ ]=11	9+[ ]=13	9+[ ]=15	9+[ ]=14
9+[ ]=16	9+[ ]=15	9+[ ]=9	9+[ ]=19
9+[ ]=12	9+[ ]=13	9+[ ]=14	9+[ ]=16
9+[ ]=15	9+[ ]=12	9+[ ]=10	9+[ ]=9
9+[ ]=13	9+[ ]=9	9+[ ]=17	9+[ ]=11
9+[ ]=14	9+[ ]=19	9+[ ]=15	9+[ ]=16
9+[ ]=17	9+[ ]=13	9+[ ]=18	9+[ ]=15
9+[ ]=12	9+[ ]=16	9+[ ]=11	9+[ ]=19
9+[ ]=10	9+[ ]=13	9+[ ]=15	9+[ ]=9

Lesson 105*   Plus Doubles

Facts: N + N (1–5)

Name: _______________________________

4+4=	2+2=	3+3=	1+1=
3+3=	4+4=	2+2=	5+5=
0+0=	1+1=	5+5=	4+4=
2+2=	5+5=	4+4=	3+3=
5+5=	1+1=	3+3=	0+0=
1+1=	0+0=	2+2=	4+4=
5+5=	3+3=	1+1=	2+2=
4+4=	2+2=	0+0=	1+1=
0+0=	5+5=	4+4=	3+3=
1+1=	3+3=	2+2=	0+0=
2+2=	5+5=	1+1=	3+3=

Lesson 106*   Plus Doubles

Facts: N + N (5–10)

Name: _______________________

7+7=	9+9=	8+8=	10+10=
8+8=	7+7=	9+9=	6+6=
5+5=	10+10=	6+6=	7+7=
9+9=	6+6=	7+7=	8+8=
6+6=	10+10=	8+8=	5+5=
10+10=	5+5=	9+9=	7+7=
6+6=	8+8=	10+10=	9+9=
7+7=	9+9=	5+5=	10+10=
9+9=	6+6=	7+7=	8+8=
10+10=	8+8=	9+9=	5+5=
6+6=	5+5=	10+10=	8+8=

Lesson 107   Plus Doubles

Facts: N+N (1–10)

Name: _______________________

4+4=	10+10=	9+9=	7+7=
3+3=	4+4=	2+2=	10+10=
6+6=	7+7=	10+10=	9+9=
2+2=	8+8=	9+9=	6+6=
7+7=	1+1=	8+8=	4+4=
10+10=	6+6=	7+7=	1+1=
8+8=	9+9=	6+6=	8+8=
1+1=	2+2=	4+4=	9+9=
6+6=	5+5=	1+1=	2+2=
5+5=	7+7=	9+9=	6+6=
9+9=	3+3=	2+2=	5+5=

Lesson 108   Plus Mix: 1s, 2s, 3s

Facts: +1, +2, +3 mix: a

Name: _______________________________

9+3=	11+3=	5+3=	7+3=
5+2=	4+1=	12+2=	8+3=
10+2=	9+2=	11+3=	12+2=
8+3=	9+3=	2+3=	6+2=
6+2=	7+3=	11+2=	9+3=
9+2=	8+3=	2+1=	5+2=
5+1=	11+3=	7+2=	10+3=
8+3=	9+1=	10+2=	8+1=
11+2=	7+3=	12+2=	5+2=
10+1=	6+2=	5+3=	11+2=
5+2=	3+3=	9+3=	12+2=

Lesson 109   9 Plus Algebra

Facts: alg 9+: b

Name: _______________________

9+[ ]=10	9+[ ]=16	9+[ ]=17	9+[ ]=18
9+[ ]=19	9+[ ]=14	9+[ ]=18	9+[ ]=10
9+[ ]=11	9+[ ]=13	9+[ ]=15	9+[ ]=14
9+[ ]=16	9+[ ]=15	9+[ ]=9	9+[ ]=19
9+[ ]=12	9+[ ]=13	9+[ ]=14	9+[ ]=16
9+[ ]=15	9+[ ]=12	9+[ ]=10	9+[ ]=9
9+[ ]=13	9+[ ]=9	9+[ ]=17	9+[ ]=11
9+[ ]=14	9+[ ]=19	9+[ ]=15	9+[ ]=16
9+[ ]=17	9+[ ]=13	9+[ ]=18	9+[ ]=15
9+[ ]=12	9+[ ]=16	9+[ ]=11	9+[ ]=19
9+[ ]=10	9+[ ]=13	9+[ ]=15	9+[ ]=9

Lesson 110 T Plus Doubles

Facts: N + N (1–5)

Name: _______________________

4+4=	2+2=	3+3=	1+1=
3+3=	4+4=	2+2=	5+5=
0+0=	1+1=	5+5=	4+4=
2+2=	5+5=	4+4=	3+3=
5+5=	1+1=	3+3=	0+0=
1+1=	0+0=	2+2=	4+4=
5+5=	3+3=	1+1=	2+2=
4+4=	2+2=	0+0=	1+1=
0+0=	5+5=	4+4=	3+3=
1+1=	3+3=	2+2=	0+0=
2+2=	5+5=	1+1=	3+3=

Lesson 111*   Minus Mix: 1s, 2s, 9s, 10s

Name: _______________________________

11-9=	12-10=	7-2=	5-1=
4-2=	8-1=	15-10=	12-9=
9-9=	18-9=	10-10=	9-2=
3-1=	6-2=	8-1=	13-9=
17-2=	19-9=	6-2=	11-10=
14-1=	5-2=	8-1=	10-9=
11-9=	10-1=	5-2=	7-1=
9-9=	8-2=	4-1=	10-2=
10-9=	15-9=	11-2=	12-10=
6-1=	11-10=	10-2=	5-2=
3-1=	12-9=	5-1=	9-2=

Lesson 112 **T**   Minus Mix: 1s, 2s, 9s, 10s

Name: _______________________________

5 − 1	7 − 2	12 − 9	8 − 2	12 − 2
10 − 2	4 − 1	10 −10	7 − 2	10 − 1
11 − 1	10 − 9	13 −10	9 − 9	6 − 2
5 − 2	6 − 1	11 − 9	12 − 2	8 − 1
10 − 2	10 −10	9 − 1	5 − 1	12 − 9
4 − 2	4 − 1	17 −10	9 − 9	11 − 2
10 − 9	7 − 2	10 − 1	11 −10	9 − 1
3 − 1	11 − 2	5 − 1	12 − 2	15 −10
5 − 2	9 − 2	13 − 2	8 − 1	7 − 2

Lesson 113*   2-Digit-2 Plus

Facts: 2 digit 2+

Name: _______________________________

52+9=	82+8=	82+4=	52+10=
52+3=	82+4=	52+7=	52+5=
52+2=	82+6=	52+8=	52+1=
52+7=	82+8=	52+9=	52+4=
82+0=	82+10=	52+5=	52+2=
82+10=	82+2=	52+3=	32+6=
72+3=	32+7=	32+1=	72+2=
72+4=	32+9=	32+0=	72+3=
72+2=	72+3=	32+10=	72+8=
72+7=	72+10=	72+3=	32+2=
32+5=	32+0=	72+6=	72+9=

Lesson 114   2-Digit-2 Plus

Name: _______________________

32+8=	32+5=	32+4=	82+10=
72+1=	72+4=	32+0=	82+4=
72+5=	32+3=	32+7=	82+8=
72+4=	32+5=	32+9=	82+6=
72+6=	32+0=	72+10=	82+7=
32+3=	32+2=	72+3=	82+5=
32+8=	32+3=	72+9=	52+7=
32+6=	32+1=	82+3=	52+2=
72+10=	32+4=	82+7=	52+6=
72+2=	32+9=	82+5=	92+2=
32+4=	32+0=	82+9=	92+7=

Lesson 115   9 Plus Algebra

Name: _______________________________

9+[ ]=10   9+[ ]=16   9+[ ]=17   9+[ ]=18

9+[ ]=19   9+[ ]=14   9+[ ]=18   9+[ ]=10

9+[ ]=11   9+[ ]=13   9+[ ]=15   9+[ ]=14

9+[ ]=16   9+[ ]=15   9+[ ]=9   9+[ ]=19

9+[ ]=12   9+[ ]=13   9+[ ]=14   9+[ ]=16

9+[ ]=15   9+[ ]=12   9+[ ]=10   9+[ ]=9

9+[ ]=13   9+[ ]=9   9+[ ]=17   9+[ ]=11

9+[ ]=14   9+[ ]=19   9+[ ]=15   9+[ ]=16

9+[ ]=17   9+[ ]=13   9+[ ]=18   9+[ ]=15

9+[ ]=12   9+[ ]=16   9+[ ]=11   9+[ ]=19

9+[ ]=10   9+[ ]=13   9+[ ]=15   9+[ ]=9

Lesson 116 T   2-Digit-2 Plus

Facts: 2 digit 2+

Name: ______________________

52+9=	82+8=	82+4=	52+10=
52+3=	82+4=	52+7=	52+5=
52+2=	82+6=	52+8=	52+1=
52+7=	82+8=	52+9=	52+4=
82+0=	82+10=	52+5=	52+2=
82+10=	82+2=	52+3=	32+6=
72+3=	32+7=	32+1=	72+2=
72+4=	32+9=	32+0=	72+3=
72+2=	72+3=	32+10=	72+8=
72+7=	72+10=	72+3=	32+2=
32+5=	32+0=	72+6=	72+9=

Lesson 117   Plus Doubles

Name: _______________________

4+4=	2+2=	3+3=	1+1=
3+3=	4+4=	2+2=	5+5=
0+0=	1+1=	5+5=	4+4=
2+2=	5+5=	4+4=	3+3=
5+5=	1+1=	3+3=	0+0=
1+1=	0+0=	2+2=	4+4=
5+5=	3+3=	1+1=	2+2=
4+4=	2+2=	0+0=	1+1=
0+0=	5+5=	4+4=	3+3=
1+1=	3+3=	2+2=	0+0=
2+2=	5+5=	1+1=	3+3=

Facts: +9, 9+: a

Name: _______________________

4+9=	2+9=	9+9 =	9+4=
1+9=	10+9=	9+3=	9+90=
5+9=	6+9=	9+6 =	9+8=
3+9=	5+9=	9+5=	3+9=
6+9=	9+9=	9+2=	9+1=
8+9=	4+9=	9+7=	9+10=
2+9=	7+9=	9+4=	9+5=
10+9=	3+9=	9+6=	9+3=
5+9=	8+9=	9+7=	9+8=
1+9=	10+9=	3+9=	4+9=
2+9=	9+3=	6+9=	9+1=

Lesson 119 9 Plus Algebra

Facts: alg 9+: a

Name: _______________________

9+[ ]=11	9+[ ]=15	9+[ ]=9	9+[ ]=13
9+[ ]=15	9+[ ]=14	9+[ ]=13	9+[ ]=16
9+[ ]=9	9+[ ]=16	9+[ ]=18	9+[ ]=14
9+[ ]=17	9+[ ]=10	9+[ ]=12	9+[ ]=13
9+[ ]=19	9+[ ]=18	9+[ ]=16	9+[ ]=15
9+[ ]=11	9+[ ]=17	9+[ ]=14	9+[ ]=18
9+[ ]=9	9+[ ]=14	9+[ ]=16	9+[ ]=10
9+[ ]=13	9+[ ]=16	9+[ ]=10	9+[ ]=12
9+[ ]=17	9+[ ]=15	9+[ ]=19	9+[ ]=17
9+[ ]=14	9+[ ]=11	9+[ ]=16	9+[ ]=9
9+[ ]=16	9+[ ]=14	9+[ ]=11	9+[ ]=15

Lesson 120   2-Digit-2 Plus

Name: _______________________

32+8=	32+5=	32+4=	82+10=
72+1=	72+4=	32+0=	82+4=
72+5=	32+3=	32+7=	82+8=
72+4=	32+5=	32+9=	82+6=
72+6=	32+0=	72+10=	82+7=
32+3=	32+2=	72+3=	82+5=
32+8=	32+3=	72+9=	52+7=
32+6=	32+1=	82+3=	52+2=
72+10=	32+4=	82+7=	52+6=
72+2=	32+9=	82+5=	92+2=
32+4=	32+0=	82+9=	92+7=

Answer Key

Answer Key

Lesson 21*   Plus 0s and 1s

Name: ___________________________

1+1=2	1+0=1	1+2=3	1+4=5
1+3=4	1+5=6	1+1=2	1+5=6
1+4=5	1+1=2	1+2=3	0+3=3
1+2=3	1+3=4	1+4=5	1+2=3
1+1=2	1+0=1	1+5=6	1+4=5
1+5=6	1+2=3	0+1=1	1+0=1
1+3=4	1+0=1	1+2=3	1+3=4
1+2=3	1+5=6	1+4=5	1+5=6
1+3=4	1+4=5	1+0=1	1+1=2
1+5=6	1+3=4	1+1=2	1+2=3
0+4=4	1+1=2	1+5=6	1+0=1

Lesson 22   Plus 0s and 1s

Name: ___________________________

1+3=4	1+4=5	1+5=6	1+1=2
1+2=3	1+0=1	1+1=2	1+3=4
1+0=1	1+3=4	0+1=1	1+0=1
0+5=5	1+2=3	1+1=2	0+4=4
1+2=3	1+5=6	1+0=1	1+2=3
1+4=5	1+3=4	1+1=2	1+3=4
1+2=3	1+1=2	0+3=3	1+0=1
1+3=4	1+4=5	1+2=3	1+1=2
1+2=3	1+3=4	1+4=5	1+5=6
1+1=2	1+0=1	1+2=3	1+3=4
1+5=6	1+4=5	1+0=1	0+2=2

Lesson 23*   Plus 0s and 1s

Name: _______________________

9+1=10	4+1=5	6+1=7	0+5=5
3+0=3	5+1=6	3+1=4	7+0=7
1+1=2	7+1=8	10+1=11	1+0=1
1+0=1	5+0=5	2+0=2	6+1=7
4+1=5	7+0=7	8+0=8	3+1=4
5+1=6	1+0=1	9+1=10	10+1=11
7+0=7	6+1=7	5+1=6	2+0=2
6+1=7	3+1=4	1+0=1	7+0=7
3+1=4	0+4=4	4+1=5	8+0=8
10+1=11	2+0=2	5+1=6	9+1=10
2+0=2	8+0=8	7+0=7	8+0=8

Lesson 24   Plus 0s and 1s

Name: ______________________

5+1=6	4+1=5	1+1=2	15+1=16
7+0=7	0+3=3	2+1=3	17+0=17
6+1=7	7+1=8	10+1=11	11+0=11
9+1=10	5+1=6	3+0=3	16+1=17
3+0=3	7+0=7	6+0=6	13+1=14
1+1=2	1+0=1	8+1=9	10+1=11
1+0=1	6+1=7	7+1=8	12+0=12
4+1=5	3+1=4	1+1=2	17+0=17
10+1=11	4+1=5	4+0=4	18+0=18
2+0=2	0+10=10	4+1=5	9+1=10
1+1=2	8+0=8	9+1=10	18+0=18

Lesson 25   Plus 0s and 1s

Facts: +1, 1+, 0 mix: a

Name: ___________________________

7+1=8	20+1=21	1+4=5	1+5=6
4+1=5	50+1=51	0+12=12	1+11=12
9+1=10	60+1=61	1+6=7	1+8=9
3+1=4	30+1=31	1+9=10	1+4=5
6+0=6	90+1=91	1+3=4	1+1=2
8+1=9	40+1=41	1+7=8	1+16=17
2+1=3	3+0=3	1+4=5	1+5=6
11+1=12	7+1=8	1+6=7	1+11=12
1+1=2	4+1=5	0+10=10	1+9=10
1+16=17	9+1=10	50+1=51	1+4=5
30+1=31	1+3=4	1+16=17	1+0=1

Lesson 26   Plus 0s and 1s

Name: _______________________________

1+5=6	6+1=7	60+1=61	6+1=7
1+11=12	3+0=3	30+1=31	8+1=9
1+8=9	8+1=9	90+1=91	2+1=3
1+4=5	1+3=4	40+0=40	11+1=12
1+1=2	40+1=41	1+15=16	15+1=16
8+1=9	12+1=13	1+80=81	1+16=17
15+1=16	0+5=5	1+40=41	7+1=8
1+9=10	1+11=12	1+50=51	4+1=5
8+1=9	1+6=7	1+18=19	9+1=10
1+9=10	1+4=5	1+8=9	3+1=4
1+1=2	12+0=12	1+50=51	6+1=7

Lesson 27   Plus 0s and 1s

Facts: +1, +0: a

Name: ___________________

9+1=10	4+1=5	6+1=7	0+5=5
3+0=3	5+1=6	3+1=4	7+0=7
1+1=2	7+1=8	10+1=11	1+0=1
1+0=1	5+0=5	2+0=2	6+1=7
4+1=5	7+0=7	8+0=8	3+1=4
5+1=6	1+0=1	9+1=10	10+1=11
7+0=7	6+1=7	5+1=6	2+0=2
6+1=7	3+1=4	1+0=1	7+0=7
3+1=4	0+4=4	4+1=5	8+0=8
10+1=10	2+0=2	5+1=6	9+1=10
2+0=2	8+0=8	7+0=7	8+0=8

Lesson 28   Plus 0s and 1s

Name: _______________________

7+1=8	20+1=21	1+4=5	1+5=6
4+1=5	50+1=51	0+12=12	1+11=12
9+1=10	60+1=61	1+6=7	1+8=9
3+1=4	30+1=31	1+9=10	1+4=5
6+0=6	90+1=91	1+3=4	1+1=2
8+1=9	40+1=41	1+7=8	1+16=17
2+1=3	3+0=3	1+4=5	1+5=6
11+1=12	7+1=8	1+6=7	1+11=12
1+1=2	4+1=5	0+10=10	1+9=10
1+16=17	9+1=10	50+1=51	1+4=5
30+1=31	1+3=4	1+16=17	1+0=1

Lesson 29   Plus 0s and 1s

Name: ___________________________

5+1=6	4+1=5	1+1=2	15+1=16
7+0=7	0+3=3	2+1=3	17+0=17
6+1=7	7+1=8	10+1=11	11+0=11
9+1=10	5+1=6	3+0=3	16+1=17
3+0=3	7+0=7	6+0=6	13+1=14
1+1=2	1+0=1	8+1=9	10+1=11
1+0=1	6+1=7	7+1=8	12+0=12
4+1=5	3+1=4	1+1=2	17+0=17
10+1=11	4+1=5	4+0=4	18+0=18
2+0=2	0+10=10	4+1=5	9+1=10
1+1=2	8+0=8	9+1=10	18+0=18

Lesson 30 T   Plus 0s and 1s

Name: ___________________________

1+5=6	6+1=7	60+1=61	6+1=7
1+11=12	3+0=3	30+1=31	8+1=9
1+8=9	8+1=9	90+1=91	2+1=3
1+4=5	1+3=4	40+0=40	11+1=12
1+1=2	40+1=41	1+15=16	15+1=16
8+1=9	12+1=13	1+80=81	1+16=17
15+1=16	0+5=5	1+40=41	7+1=8
1+9=10	1+11=12	1+50=51	4+1=5
8+1=9	1+6=7	1+18=19	9+1=10
1+9=10	1+4=5	1+8=9	3+1=4
1+1=2	12+0=12	1+50=51	6+1=7

Lesson 31*   Minus 0s and 1s

Name: ______________________

11-1=10	2-0=2	3-1=2	5-1=4
4-1=3	8-1=7	5-1=4	11-1=10
9-0=9	1-1=0	10-1=9	9-1=8
3-1=2	6-1=5	8-1=7	2-1=1
7-1=6	3-0=3	6-1=5	10-1=9
4-1=3	5-1=4	3-1=2	2-1=1
11-1=10	10-1=9	5-0=5	7-1=6
9-1=8	8-1=7	4-1=3	10-1=9
2-1=1	5-1=4	11-1=10	1-0=1
6-0=6	11-1=10	10-1=9	5-1=4
3-1=2	1-1=0	5-1=4	9-1=8

Lesson 32 T   Minus 0s and 1s

Name: _______________________

5-1=4	7-1=6	11-0=11	4-1=3
10-1=9	4-1=3	2-1=1	7-1=6
8-0=8	9-1=8	4-1=3	10-1=9
11-1=10	10-1=9	2-1=1	3-1=2
5-1=4	6-0=6	11-1=10	1-1=0
10-1=9	3-1=2	9-1=8	5-1=4
4-1=3	4-0=4	7-1=6	9-0=9
2-1=1	7-1=6	2-1=1	11-1=10
3-1=2	11-1=10	5-0=5	1-1=0
5-1=4	9-1=8	2-1=1	8-1=7
11-0=11	10-1=9	6-1=5	4-1=3

Lesson 33*   Plus/Minus 0s and 1s

Name: _______________________

5+0=5	10-1=9	16-1=15	18+0=18
7+1=8	2+1=3	14-0=14	9-1=8
6-1=5	17+1=18	15-1=14	18+0=18
3+1=4	13-0=13	17+1=18	16-1=15
10+0=10	9-1=8	12-0=12	13-1=12
2+0=2	15-0=15	7+1=8	3-0=3
1+1=2	13-1=12	1+0=1	12+1=13
5+1=6	10-1=9	6-1=5	8+0=8
3+1=4	5-0=5	3-1=2	9-1=8
9+0=9	8+0=8	4-1=3	18-1=17
1+1=2	9-1=8	2+1=3	14+1=15

Lesson 34   Plus/Minus 0s and 1s

Name: ___________________________

5+1=6	10+1=11	16-1=15	18-1=17
7-1=6	2-0=2	14+1=15	9+1=10
6-1=5	17-1=16	15+1=16	18-1=17
3+1=4	14+1=15	17-1=16	16+1=17
10+1=11	9+1=10	15+1=16	13-0=13
2-1=1	15+1=16	7-1=6	3+1=4
1+1=2	13+1=14	1-1=0	12-1=11
5+1=6	10+1=11	6-1=5	8-1=7
3-1=2	5+1=6	3+1=4	9-1=8
9+0=9	8-1=7	4+1=5	14+1=15
1+1=2	9+1=10	2-1=1	14-1=13

Name: ___________________________

5+0=5	10-1=9	16-1=15	18+0=18
7+1=8	2+1=3	14-0=14	9-1=8
6-1=5	17+1=18	15-1=14	18+0=18
3+1=4	13-0=13	17+1=18	16-1=15
10+0=10	9-1=8	12-0=12	13-1=12
2+0=2	15-0=15	7+1=8	3-0=3
1+1=2	13-1=12	1+0=1	12+1=13
5+1=6	10-1=9	6-1=5	8+0=8
3+1=4	5-0=5	3-1=2	9-1=8
9+0=9	8+0=8	4-1=3	18-1=17
1+1=2	9-1=8	2+1=3	14+1=15

Lesson 36   Minus 0s and 1s

Name: _______________________

5-1=4	7-1=6	11-0=11	4-1=3
10-1=9	4-1=3	2-1=1	7-1=6
8-0=8	9-1=8	4-1=3	10-1=9
11-1=10	10-1=9	2-1=1	3-1=2
5-1=4	6-0=6	11-1=10	1-1=0
10-1=9	3-1=2	9-1=8	5-1=4
4-1=3	4-0=4	7-1=6	9-0=9
2-1=1	7-1=6	2-1=1	11-1=10
3-1=2	11-1=10	5-0=5	1-1=0
5-1=4	9-1=8	2-1=1	8-1=7
11-0=11	10-1=9	6-1=5	4-1=3

Lesson 37   Plus 0s and 1s

Name: _______________________

1+5=6	6+1=7	60+1=61	6+1=7
1+11=12	3+0=3	30+1=31	8+1=9
1+8=9	8+1=9	90+1=91	2+1=3
1+4=5	1+3=4	40+0=40	11+1=12
1+1=2	40+1=41	1+15=16	15+1=16
8+1=9	12+1=13	1+80=81	1+16=17
15+1=16	0+5=5	1+40=41	7+1=8
1+9=10	1+11=12	1+50=51	4+1=5
8+1=9	1+6=7	1+18=19	9+1=10
1+9=10	1+4=5	1+8=9	3+1=4
1+1=2	12+0=12	1+50=51	6+1=7

Name: _______________________

11-1=10	2-0=2	3-1=2	5-1=4
4-1=3	8-1=7	5-1=4	11-1=10
9-0=9	1-1=0	10-1=9	9-1=8
3-1=2	6-1=5	8-1=7	2-1=1
7-1=6	3-0=3	6-1=5	10-1=9
4-1=3	5-1=4	3-1=2	2-1=1
11-1=10	10-1=9	5-0=5	7-1=6
9-1=8	8-1=7	4-1=3	10-1=9
2-1=1	5-1=4	11-1=10	1-0=1
6-0=6	11-1=10	10-1=9	5-1=4
3-1=2	1-1=0	5-1=4	9-1=8

Lesson 39 T   Plus/Minus 0s and 1s Facts: + − 0, 1 mix: b

Name: _______________________________

5+1=6	10+1=11	16-1=15	18-1=17
7-1=6	2-0=2	14+1=15	9+1=10
6-1=5	17-1=16	15+1=16	18-1=17
3+1=4	14+1=15	17-1=16	16+1=17
10+1=11	9+1=10	15+1=16	13-0=13
2-1=1	15+1=16	7-1=6	3+1=4
1+1=2	13+1=14	1-1=0	12-1=11
5+1=6	10+1=11	6-1=5	8-1=7
3-1=2	5+1=6	3+1=4	9-1=8
9+0=9	8-1=7	4+1=5	14+1=15
1+1=2	9+1=10	2-1=1	14-1=13

Lesson 40   Plus 0s and 1s

Facts: +1, 1+, 0 mix: a

Name: _______________________________

7+1=8	20+1=21	1+4=5	1+5=6
4+1=5	50+1=51	0+12=12	1+11=12
9+1=10	60+1=61	1+6=7	1+8=9
3+1=4	30+1=31	1+9=10	1+4=5
6+0=6	90+1=91	1+3=4	1+1=2
8+1=9	40+1=41	1+7=8	1+16=17
2+1=3	3+0=3	1+4=5	1+5=6
11+1=12	7+1=8	1+6=7	1+11=12
1+1=2	4+1=5	0+10=10	1+9=10
1+16=17	9+1=10	50+1=51	1+4=5
30+1=31	1+3=4	1+16=17	1+0=1

Lesson 41*   Plus 2s

Name: _______________________________

2+3=5	2+4=6	2+1=3	2+5=7
2+2=4	2+5=7	2+3=5	2+1=3
2+0=2	2+3=5	2+0=2	2+2=4
2+5=7	2+2=4	2+1=3	2+0=2
2+2=4	2+5=7	2+2=4	2+5=7
2+4=6	2+3=5	2+4=6	2+2=4
2+5=7	2+1=3	2+0=2	2+5=7
2+3=5	2+4=6	2+2=4	2+0=2
2+2=4	2+3=5	2+5=7	2+1=3
2+1=3	2+0=2	2+3=5	2+4=6
2+5=7	2+1=3	2+2=4	2+0=2

Lesson 42   Plus 2s

Name: _______________________

2+1=3	2+0=2	2+2=4	2+4=6
2+3=5	2+5=7	2+1=3	2+5=7
2+4=6	2+1=3	2+2=4	2+3=5
2+2=4	2+0=2	2+4=6	2+2=4
2+1=3	2+3=5	2+5=7	2+4=6
2+5=7	2+2=4	2+1=3	2+0=2
2+3=5	2+0=2	2+2=4	2+3=5
2+2=4	2+5=7	2+4=6	2+5=7
2+3=5	2+4=6	2+0=2	2+1=3
2+5=7	2+3=5	2+1=3	2+2=4
2+4=6	2+5=7	2+3=5	2+0=2

Lesson 43*   Plus 2s

Name: ________________________

2+9=11   2+8=10   2+4=6   2+10=12

2+3=5   2+4=6   2+7=9   2+5=7

2+2=4   2+6=8   2+8=10   2+1=3

2+7=9   2+8=10   2+9=11   2+4=6

2+0=2   2+10=12   2+5=7   2+2=4

2+10=12   2+2=4   2+3=5   2+6=8

2+3=5   2+7=9   2+1=3   2+2=4

2+4=6   2+9=11   2+0=2   2+3=5

2+2=4   2+3=5   2+10=12   2+8=10

2+7=9   2+10=12   2+3=5   2+2=4

2+5=7   2+0=2   2+6=8   2+9=11

Lesson 44   Plus 2s

Name: _______________________

2+8=10	2+5=7	2+4=6	2+10=12
2+1=3	2+4=6	2+0=2	2+4=6
2+5=7	2+3=5	2+7=9	2+8=10
2+4=6	2+5=7	2+9=11	2+6=8
2+6=8	2+0=2	2+10=12	2+7=9
2+3=5	2+2=4	2+3=5	2+5=7
2+8=10	2+3=5	2+9=11	2+7=9
2+6=8	2+1=3	2+3=5	2+2=4
2+10=12	2+4=6	2+7=9	2+6=8
2+2=4	2+9=11	2+5=7	2+7=9
2+4=6	2+0=2	2+9=11	2+2=4

Name: ___________________________

4+2=6	0+2=2	6+2=8	5+2=7
8+2=10	2+2=4	4+2=6	8+2=10
9+2=11	5+2=7	7+2=9	6+2=8
4+2=6	8+2=10	3+2=5	2+2=4
9+2=11	1+2=3	4+2=6	6+2=8
10+2=12	3+2=5	0+2=2	5+2=7
8+2=10	4+2=6	9+2=11	7+2=9
2+2=4	3+2=5	5+2=7	10+2=12
1+2=3	7+2=9	6+2=8	4+2=6
2+2=4	4+2=6	3+2=5	6+2=8
6+2=8	3+2=5	1+2=3	10+2=12

Name: ___________________________

2+2=4	9+2=11	7+2=9	6+2=8
5+2=7	6+2=8	4+2=6	8+2=10
2+2=4	3+2=5	10+2=12	2+2=4
9+2=11	8+2=10	7+2=9	5+2=7
4+2=6	1+2=3	8+2=10	6+2=8
5+2=7	10+2=12	6+2=8	7+2=9
8+2=10	2+2=4	5+2=7	4+2=6
5+2=7	6+2=8	8+2=10	9+2=11
6+2=8	1+2=3	0+2=2	4+2=6
8+2=10	6+2=8	10+2=12	3+2=5
5+2=7	0+2=2	2+2=4	7+2=9

Name: ___________________________

2+6=8	2+3=5	2+4=6	2+5=7
2+8=10	2+7=9	2+20=22	2+8=10
2+4=6	2+9=11	2+7=9	2+3=5
1+2=3	4+2=6	2+6=8	2+1=3
2+2=4	20+2=22	7+2=9	5+2=7
2+5=7	60+2=62	4+2=6	2+4=6
2+6=8	30+2=32	9+2=11	2+8=10
60+2=62	90+2=92	3+2=5	2+9=11
2+7=9	9+2=11	2+6=8	90+2=92
2+20=22	1+2=3	2+10=12	2+8=10
2+5=7	2+9=11	50+2=52	2+5=7

Lesson 48   Plus 2s

Name: ___________________________

2+1=3	2+3=5	30+2=32	2+4=6
2+8=10	2+7=9	1+2=3	2+6=8
2+4=6	2+10=12	40+2=42	2+2=4
2+40=42	2+80=82	8+2=10	20+2=22
2+10=12	2+3=5	2+40=42	2+5=7
2+4=6	2+40=42	2+50=52	2+1=3
8+2=10	2+2=4	2+8=10	2+9=11
2+2=4	2+5=7	2+3=5	2+90=92
2+6=8	2+2=4	2+80=82	1+2=3
2+7=9	2+9=11	2+5=7	5+2=7
2+8=10	2+5=7	2+2=4	2+6=8

Lesson 49   Plus 2s

Name: _______________________

6+2=8	3+2=5	2+4=6	2+5=7
8+2=10	7+2=9	2+20=22	2+2=4
2+4=6	40+2=42	50+2=52	2+3=5
1+2=3	4+2=6	2+6=8	2+1=3
2+2=4	20+2=22	7+2=9	2+2=4
2+5=7	60+2=62	4+2=6	2+4=6
2+6=8	30+2=32	9+2=11	2+8=10
60+2=62	90+2=92	3+2=5	2+9=11
2+7=9	9+2=11	2+6=8	90+2=92
2+20=22	1+2=3	2+20=22	2+8=10
2+5=7	2+9=11	50+2=52	2+5=7

Lesson 50 T   Plus 2s

Name: ________________________________

2+1=3	2+3=5	30+2=32	2+4=6
2+8=10	2+7=9	1+2=3	2+6=8
2+4=6	2+10=12	40+2=42	2+2=4
2+40=42	2+80=82	9+2=11	20+2=22
2+10=12	2+3=5	2+40=42	2+5=7
2+4=6	2+40=42	2+50=52	2+1=3
8+2=10	2+2=4	2+8=10	2+9=11
2+2=4	5+2=7	3+2=5	7+2=9
6+2=8	2+2=4	4+2=6	1+2=3
2+7=9	2+9=11	2+5=7	5+2=7
2+8=10	2+5=7	2+2=4	2+6=8

Name: _______________________

11-2=9	2-2=0	3-2=1	5-2=3
4-2=2	8-2=6	5-2=3	12-2=10
7-2=5	12-2=10	10-2=8	9-2=7
3-2=1	6-2=4	8-2=6	2-2=0
7-2=5	3-2=1	6-2=4	11-2=9
4-2=2	5-2=3	3-2=1	2-2=0
11-2=9	10-2=8	5-2=3	7-2=5
9-2=7	8-2=6	4-2=2	10-2=8
2-2=0	5-2=3	11-2=9	12-2=10
6-2=4	11-2=9	10-2=8	5-2=3
3-2=1	12-2=10	5-2=3	9-2=7

Name: _______________________

5-2=3	7-2=5	12-2=10	4-2=2
10-2=8	4-2=2	2-2=0	7-2=5
8-2=6	9-2=7	4-2=2	10-2=8
11-2=9	10-2=8	2-2=0	3-2=1
5-2=3	6-2=4	11-2=9	12-2=10
10-2=8	3-2=1	9-2=7	5-2=3
4-2=2	6-2=4	7-2=5	9-2=7
2-2=0	7-2=5	2-2=0	11-2=9
3-2=1	11-2=9	5-2=3	12-2=10
5-2=3	9-2=7	2-2=0	8-2=6
12-2=10	10-2=8	6-2=4	4-2=2

Lesson 53   Plus Mix: 0s, 1s, 2s   Facts: +2, +1, +0 mix: a

Name: _______________________

3+2=5	11+1=12	5+1=6	2+2=4
5+2=7	4+0=4	12+2=14	8+1=9
10+2=12	9+2=11	9+1=10	12+2=14
8+1=9	3+2=5	2+1=3	3+2=5
6+2=8	7+1=8	11+2=13	6+2=8
3+2=5	4+2=6	2+0=2	5+2=7
5+0=5	11+1=12	7+2=9	10+1=11
4+2=6	9+0=9	10+2=12	8+0=8
11+2=13	2+2=4	12+2=14	5+2=7
10+0=10	6+2=8	5+1=6	11+2=13
8+1=9	9+0=9	3+2=5	6+1=7

Lesson 54 T   Plus Mix: 0s, 1s, 2s

Facts: +2, +1, +0 mix: b

5+2=7   3+1=4   9+1=10   1+2=3

2+2=4   5+2=7   4+0=4   7+1=8

2+1=3   10+2=12   7+2=9   4+2=6

4+1=5   8+2=10   10+1=11   9+2=11

2+2=4   11+2=13   3+2=5   10+0=10

11+0=11   5+1=6   12+2=14   6+2=8

9+2=11   10+0=10   5+2=7   3+2=5

7+2=9   4+2=6   9+2=11   4+2=6

4+2=6   2+2=4   11+1=12   7+0=7

9+2=11   9+1=10   9+0=9   6+2=8

7+2=9   10+1=11   4+0=4   3+2=5

Name: _______________________________

11-1=10	2-2=0	3-2=1	5-1=4
4-0=4	8-1=7	5-2=3	12-2=10
9-2=7	12-2=10	10-2=8	9-1=8
3-2=1	6-2=4	8-1=7	2-1=1
7-1=6	3-2=1	6-2=4	11-2=9
4-2=2	5-2=3	3-2=1	2-0=2
11-1=10	10-1=9	5-0=5	7-2=5
9-0=9	8-0=8	4-2=2	10-2=8
2-2=0	5-2=3	11-2=9	12-2=10
6-2=4	11-2=9	10-0=10	5-1=4
3-1=2	12-2=10	5-2=3	9-1=8

Lesson 56   Minus Mix: 0s, 1s, 2s

Name: ___________________________

5-2=3	7-1=6	12-2=10	4-0=4
10-2=8	4-1=3	2-1=1	7-2=5
8-2=6	9-2=7	4-1=3	10-1=9
11-2=9	10-0=10	2-2=0	3-2=1
5-1=4	6-2=4	11-0=11	12-2=10
10-0=10	3-2=1	9-2=7	5-2=3
4-1=3	4-2=2	7-2=5	9-2=7
2-2=0	7-0=7	2-2=0	11-1=10
3-0=3	11-2=9	5-1=4	12-2=10
5-1=4	9-2=7	2-2=0	8-1=7
12-2=10	10-1=9	6-2=4	4-1=3

Lesson 57 T   Minus Mix: 0s, 1s, 2s   Facts: −2, −1, −0 mix: a

Name: _______________________________

11-1=10	2-2=0	3-2=1	5-1=4
4-0=4	8-1=7	5-2=3	12-2=10
9-2=7	12-2=10	10-2=8	9-1=8
3-2=1	6-2=4	8-1=7	2-1=1
7-1=6	3-2=1	6-2=4	11-2=9
4-2=2	5-2=3	3-2=1	2-0=2
11-1=10	10-1=9	5-0=5	7-2=5
9-0=9	8-0=8	4-2=2	10-2=8
2-2=0	5-2=3	11-2=9	12-2=10
6-2=4	11-2=9	10-0=10	5-1=4
3-1=2	12-2=10	5-2=3	9-1=8

Lesson 58　　Plus Mix: 0s, 1s, 2s

Facts:　+2, +1, +0 mix:　b

Name: _______________________

5+2=7	3+1=4	9+1=10	1+2=3
2+2=4	5+2=7	4+0=4	7+1=8
2+1=3	10+2=12	7+2=9	4+2=6
4+1=5	8+2=10	10+1=11	9+2=11
2+2=4	11+2=13	3+2=5	10+0=10
11+0=11	5+1=6	12+2=14	6+2=8
9+2=11	10+0=10	5+2=7	3+2=5
7+2=9	4+2=6	9+2=11	4+2=6
4+2=6	2+2=4	11+1=12	7+0=7
9+2=11	9+1=10	9+0=9	6+2=8
7+2=9	10+1=11	4+0=4	3+2=5

Lesson 59   Plus Mix: 0s, 1s, 2s

Facts: +2, +1, +0 mix: a

Name: _______________________

3+2=5	11+1=12	5+1=6	2+2=4
5+2=7	4+0=4	12+2=14	8+1=9
10+2=12	9+2=11	9+1=10	12+2=14
8+1=9	3+2=5	2+1=3	3+2=5
6+2=8	7+1=8	11+2=13	6+2=8
3+2=5	4+2=6	2+0=2	5+2=7
5+0=5	11+1=12	7+2=9	10+1=11
4+2=6	9+0=9	10+2=12	8+0=8
11+2=13	2+2=4	12+2=14	5+2=7
10+0=10	6+2=8	5+1=6	11+2=13
8+1=9	9+0=9	3+2=5	6+1=7

Lesson 60   Minus Mix: 0s, 1s, 2s   Facts: −2, −1, −0 mix: b

5-2=3	7-1=6	12-2=10	4-0=4
10-2=8	4-1=3	2-1=1	7-2=5
8-2=6	9-2=7	4-1=3	10-1=9
11-2=9	10-0=10	2-2=0	3-2=1
5-1=4	6-2=4	11-0=11	12-2=10
10-0=10	3-2=1	9-2=7	5-2=3
4-1=3	4-2=2	7-2=5	9-2=7
2-2=0	7-0=7	2-2=0	11-1=10
3-0=3	11-2=9	5-1=4	12-2=10
5-1=4	9-2=7	2-2=0	8-1=7
12-2=10	10-1=9	6-2=4	4-1=3

Lesson 61*   Plus 10s

Facts: +10: a

Name: _______________________________

9+10=19   4+10=14   6+10=16   5+10=15

3+10=13   5+10=15   3+10=13   7+10=17

1+10=11   7+10=17   10+10=20   1+10=11

10+10=20   5+10=15   2+10=12   6+10=16

4+10=14   7+10=17   8+10=18   3+10=13

5+10=15   1+10=11   9+10=19   10+10=20

7+10=17   6+10=16   5+10=15   2+10=12

6+10=16   3+10=13   1+10=11   7+10=17

3+10=13   4+10=14   2+10=12   8+10=18

10+10=20   2+10=12   5+10=15   9+10=19

2+10=12   8+10=18   7+10=17   8+10=18

Lesson 62   Plus 10s

Name: _______________________

10+5=15	10+6=16	10+4=14	10+5=15
10+9=19	10+7=17	10+10=20	10+7=17
10+8=18	10+3=13	10+6=16	10+8=18
10+4=14	10+8=18	10+9=19	10+4=14
10+7=17	10+2=12	10+3=13	10+5=15
10+6=16	10+4=14	10+7=17	10+10=20
10+9=19	10+6=16	10+10=20	10+5=15
10+6=16	10+7=17	10+5=15	10+9=19
10+6=16	10+4=14	10+8=18	10+3=13
10+4=14	10+6=16	10+4=14	10+2=12
10+7=17	10+10=20	10+5=15	10+1=11

Lesson 63   Plus 10s

Name: _______________________________

7+10=17	2+10=12	10+4=14	10+5=15
4+10=14	6+10=16	10+2=12	10+10=20
9+10=19	3+10=13	5+10=15	10+3=13
3+10=13	9+10=19	10+6=16	10+10=20
6+10=16	3+10=13	10+9=19	10+2=12
8+10=18	7+10=17	10+3=13	10+4=14
2+10=12	4+10=14	10+7=17	10+5=15
10+10=20	3+10=13	10+4=14	10+6=16
0+10=10	9+10=19	10+7=17	6+10=16
10+6=16	10+10=20	10+3=13	10+7=17
1+10=11	10+5=15	8+10=18	10+2=12

Lesson 64 T   Plus 10s

Name: _______________________________

10+5=15	10+8=18	5+10=15	10+8=18
10+10=20	10+3=13	3+10=13	10+9=19
10+8=18	10+4=14	10+2=12	9+10=19
10+3=13	10+10=20	4+10=14	8+10=18
10+7=17	10+5=15	8+10=18	10+9=19
10+5=15	10+10=20	10+2=12	10+4=14
10+0=10	10+7=17	10+5=15	10+6=16
8+10=18	10+8=18	10+6=16	10+10=20
10+5=15	4+10=14	10+3=13	6+10=16
10+6=16	10+2=12	10+8=18	10+5=15
10+7=17	6+10=16	3+10=13	10+10=20

Lesson 65   Plus Mix: 0s, 1s, 2s, 10s

Facts: +0, 1, 2, 10 mix: a

Name: _______________________________

$\begin{array}{r}10\\[-2pt]+\ 2\\\hline 12\end{array}$	$\begin{array}{r}3\\[-2pt]+\ 1\\\hline 4\end{array}$	$\begin{array}{r}5\\[-2pt]+\ 2\\\hline 7\end{array}$	$\begin{array}{r}2\\[-2pt]+\ 0\\\hline 2\end{array}$	$\begin{array}{r}6\\[-2pt]+\ 1\\\hline 7\end{array}$
$\begin{array}{r}4\\[-2pt]+\ 1\\\hline 5\end{array}$	$\begin{array}{r}2\\[-2pt]+\ 9\\\hline 11\end{array}$	$\begin{array}{r}2\\[-2pt]+\ 6\\\hline 8\end{array}$	$\begin{array}{r}1\\[-2pt]+10\\\hline 11\end{array}$	$\begin{array}{r}5\\[-2pt]+\ 0\\\hline 5\end{array}$
$\begin{array}{r}7\\[-2pt]+\ 2\\\hline 9\end{array}$	$\begin{array}{r}2\\[-2pt]+\ 6\\\hline 8\end{array}$	$\begin{array}{r}5\\[-2pt]+\ 0\\\hline 5\end{array}$	$\begin{array}{r}1\\[-2pt]+\ 3\\\hline 4\end{array}$	$\begin{array}{r}9\\[-2pt]+\ 1\\\hline 10\end{array}$
$\begin{array}{r}7\\[-2pt]+10\\\hline 17\end{array}$	$\begin{array}{r}0\\[-2pt]+\ 8\\\hline 8\end{array}$	$\begin{array}{r}6\\[-2pt]+\ 1\\\hline 7\end{array}$	$\begin{array}{r}2\\[-2pt]+\ 1\\\hline 3\end{array}$	$\begin{array}{r}2\\[-2pt]+\ 3\\\hline 5\end{array}$
$\begin{array}{r}2\\[-2pt]+\ 4\\\hline 6\end{array}$	$\begin{array}{r}7\\[-2pt]+\ 2\\\hline 9\end{array}$	$\begin{array}{r}10\\[-2pt]+\ 2\\\hline 12\end{array}$	$\begin{array}{r}3\\[-2pt]+\ 0\\\hline 3\end{array}$	$\begin{array}{r}2\\[-2pt]+\ 7\\\hline 9\end{array}$
$\begin{array}{r}3\\[-2pt]+\ 2\\\hline 5\end{array}$	$\begin{array}{r}10\\[-2pt]+\ 3\\\hline 13\end{array}$	$\begin{array}{r}9\\[-2pt]+\ 0\\\hline 9\end{array}$	$\begin{array}{r}1\\[-2pt]+\ 5\\\hline 6\end{array}$	$\begin{array}{r}8\\[-2pt]+10\\\hline 18\end{array}$
$\begin{array}{r}5\\[-2pt]+10\\\hline 15\end{array}$	$\begin{array}{r}6\\[-2pt]+\ 2\\\hline 8\end{array}$	$\begin{array}{r}9\\[-2pt]+\ 1\\\hline 10\end{array}$	$\begin{array}{r}1\\[-2pt]+\ 4\\\hline 5\end{array}$	$\begin{array}{r}3\\[-2pt]+\ 0\\\hline 3\end{array}$
$\begin{array}{r}4\\[-2pt]+\ 1\\\hline 5\end{array}$	$\begin{array}{r}7\\[-2pt]+\ 1\\\hline 8\end{array}$	$\begin{array}{r}9\\[-2pt]+\ 2\\\hline 11\end{array}$	$\begin{array}{r}0\\[-2pt]+\ 5\\\hline 5\end{array}$	$\begin{array}{r}10\\[-2pt]+\ 2\\\hline 12\end{array}$
$\begin{array}{r}9\\[-2pt]+10\\\hline 19\end{array}$	$\begin{array}{r}1\\[-2pt]+\ 1\\\hline 2\end{array}$	$\begin{array}{r}0\\[-2pt]+\ 0\\\hline 0\end{array}$	$\begin{array}{r}1\\[-2pt]+\ 3\\\hline 4\end{array}$	$\begin{array}{r}9\\[-2pt]+\ 1\\\hline 10\end{array}$

Lesson 66   Plus Mix: 0s, 1s, 2s, 10s

Name: _______________________

10+7=17	2+4=6	0+3=3	9+0=9
5+10=15	3+1=4	1+6=7	6+10=16
10+3=13	6+2=8	5+2=7	2+8=10
1+0=1	5+1=6	7+2=9	2+4=6
6+1=7	1+3=4	0+9=9	9+0=9
7+10=17	10+3=13	2+4=6	7+2=9
4+1=5	0+7=7	5+2=7	10+3=13
10+4=14	2+3=5	6+2=8	4+10=14
8+10=18	0+4=4	1+10=11	5+0=5
4+1=5	1+8=9	7+0=7	1+9=10
10+6=16	4+10=14	1+1=2	7+2=9

Lesson 67   Plus Mix: 0s, 1s, 2s, 10s

Name: _______________________

9 + 2 **11**	2 + 8 **10**	7 + 1 **8**	10 + 4 **14**	9 + 2 **11**
1 + 6 **7**	5 + 0 **5**	2 +10 **12**	2 + 6 **8**	6 + 2 **8**
0 + 7 **7**	4 + 1 **5**	0 + 2 **2**	7 +10 **17**	1 + 4 **5**
6 + 1 **7**	4 +10 **14**	10 + 1 **11**	2 + 6 **8**	0 + 6 **6**
5 + 1 **6**	0 + 8 **8**	5 + 2 **7**	10 + 8 **18**	1 + 4 **5**
7 +10 **17**	1 + 6 **7**	9 + 1 **10**	0 + 5 **5**	2 +10 **12**
10 + 0 **10**	6 + 1 **7**	7 +10 **17**	2 + 3 **5**	10 + 3 **13**
1 + 7 **8**	6 + 2 **8**	2 + 0 **2**	5 + 1 **6**	1 + 5 **6**
0 + 4 **4**	4 + 1 **5**	10 + 6 **16**	5 +10 **15**	4 + 0 **4**

Lesson 68   Plus Mix: 0s, 1s, 2s, 10s

Facts: +0, 1, 2 mix

Name: _______________________

5+1=6	11+1=12	3+2=5	2+2=4
12+2=14	4+0=4	5+2=7	8+1=9
9+1=10	9+2=11	10+2=12	12+2=14
2+1=3	3+2=5	8+1=9	6+2=8
1+2=3	7+1=8	6+2=8	3+2=5
2+0=2	4+2=6	3+2=5	5+2=7
7+2=9	11+1=12	5+0=5	10+1=11
10+2=12	9+0=9	4+2=6	8+0=8
12+2=14	2+2=4	11+2=13	5+2=7
5+1=6	6+2=8	10+0=10	11+2=13
9+1=10	3+1=4	5+2=7	12+2=14

Lesson 69 T   Plus Mix: 0s, 1s, 2s, 10s Facts: +0, 1, 2 mix

Name: _______________________

9 + 1 **10**	3 + 1 **4**	5 + 2 **7**	12 + 2 **14**	11 + 2 **13**
4 + 0 **4**	5 + 2 **7**	2 + 2 **4**	7 + 1 **8**	9 + 1 **10**
7 + 2 **9**	10 + 2 **12**	2 + 1 **3**	4 + 1 **5**	3 + 1 **4**
10 + 1 **11**	8 + 2 **10**	4 + 1 **5**	9 + 2 **11**	5 + 2 **7**
3 + 2 **5**	11 + 2 **13**	2 + 2 **4**	10 + 0 **10**	12 + 2 **14**
1 + 2 **3**	5 + 1 **6**	11 + 0 **11**	6 + 2 **8**	11 + 0 **11**
5 + 2 **7**	10 + 0 **10**	9 + 2 **11**	3 + 2 **5**	7 + 1 **8**
9 + 2 **11**	4 + 2 **6**	7 + 2 **9**	4 + 2 **6**	6 + 2 **8**
11 + 1 **12**	2 + 2 **4**	10 + 2 **12**	7 + 0 **7**	8 + 1 **9**

Lesson 70 Minus Mix: 0s, 1s, 2s

Facts: −0, 1, 2 mix

Name: _______________________

11-1=10	2-2=0	3-2=1	5-1=4
4-0=4	8-1=7	5-2=3	12-2=10
9-2=7	12-2=10	10-2=8	9-1=8
3-2=1	6-2=4	8-1=7	2-1=1
7-1=6	3-2=1	6-2=4	11-2=9
4-2=2	5-2=3	3-2=1	2-0=2
11-1=10	10-1=9	5-0=5	7-0=7
9-0=9	8-0=8	4-2=2	10-2=8
2-2=0	5-2=3	11-2=9	12-2=10
6-2=4	11-2=9	10-0=10	5-1=4
3-1=2	12-2=10	5-2=3	9-1=8

Lesson 71   Minus Mix: 0s, 1s, 2s

Facts: −0, 1, 2, mix

Name: _______________________________

5 − 2 = 3	7 − 1 = 6	12 − 2 = 10	4 − 0 = 4	4 − 2 = 2
10 − 2 = 8	4 − 1 = 3	2 − 1 = 1	7 − 2 = 5	11 − 1 = 10
8 − 2 = 6	9 − 2 = 7	4 − 1 = 3	10 − 1 = 9	8 − 0 = 8
11 − 2 = 9	10 − 0 = 10	2 − 2 = 0	3 − 2 = 1	3 − 0 = 3
5 − 1 = 4	6 − 2 = 4	11 − 0 = 11	12 − 2 = 10	10 − 1 = 9
10 − 0 = 10	3 − 2 = 1	9 − 2 = 7	5 − 2 = 3	6 − 1 = 5
4 − 2 = 2	4 − 1 = 3	7 − 2 = 5	9 − 2 = 7	5 − 2 = 3
2 − 2 = 0	7 − 0 = 7	2 − 2 = 0	11 − 1 = 10	4 − 1 = 3
3 − 0 = 3	11 − 2 = 9	5 − 1 = 4	12 − 2 = 10	9 − 2 = 7

Lesson 72   Plus 10s

Name: ___________________________

7+10=17	2+10=12	10+4=14	10+5=15
4+10=14	6+10=16	10+2=12	10+10=20
9+10=19	3+10=13	5+10=15	10+3=13
3+10=13	9+10=19	10+6=16	10+10=20
6+10=16	3+10=13	10+9=19	10+2=12
8+10=18	7+10=17	10+3=13	10+4=14
2+10=12	4+10=14	10+7=17	10+5=15
10+10=20	3+10=13	10+4=14	10+6=16
0+10=10	9+10=19	10+7=17	6+10=16
10+6=16	10+10=20	10+3=13	10+7=17
1+10=11	10+5=15	8+10=18	10+2=12

Lesson 73   Plus Mix: 0s, 1s, 2s, 10s

Facts: +0, 1, 2, 10 mix: c

Name: _______________________

9 + 2 11	2 + 8 10	7 + 1 8	10 + 4 14	9 + 2 11
1 + 6 7	5 + 0 5	2 +10 12	2 + 6 8	6 + 2 8
0 + 7 7	4 + 1 5	0 + 2 2	7 +10 17	1 + 4 5
6 +1 7	4 +10 14	10 + 1 11	2 + 6 8	0 + 6 6
5 + 1 6	0 + 8 8	5 + 2 7	10 + 8 18	1 + 4 5
7 +10 17	1 + 6 7	9 + 1 10	0 + 5 5	2 +10 12
10 + 0 10	6 + 1 7	7 +10 17	2 + 3 5	10 + 3 13
1 + 7 8	6 + 2 8	2 + 0 2	5 + 1 6	1 + 5 6
0 + 4 4	4 + 1 5	10 + 6 16	5 +10 15	4 + 0 4

Lesson 74 Minus Mix: 0s, 1s, 2s

Facts: −0, 1, 2, mix

Name: ______________________

5 − 2 = 3	7 − 1 = 6	12 − 2 = 10	4 − 0 = 4	4 − 2 = 2
10 − 2 = 8	4 − 1 = 3	2 − 1 = 1	7 − 2 = 5	11 − 1 = 10
8 − 2 = 6	9 − 2 = 7	4 − 1 = 3	10 − 1 = 9	8 − 0 = 8
11 − 2 = 9	10 − 0 = 10	2 − 2 = 0	3 − 2 = 1	3 − 0 = 3
5 − 1 = 4	6 − 2 = 4	11 − 0 = 11	12 − 2 = 10	10 − 1 = 9
10 − 0 = 10	3 − 2 = 1	9 − 2 = 7	5 − 2 = 3	6 − 1 = 5
4 − 2 = 2	4 − 1 = 3	7 − 2 = 5	9 − 2 = 7	5 − 2 = 3
2 − 2 = 0	7 − 0 = 7	2 − 2 = 0	11 − 1 = 10	4 − 1 = 3
3 − 0 = 3	11 − 2 = 9	5 − 1 = 4	12 − 2 = 10	9 − 2 = 7

Lesson 75*   Plus 9s

Facts: +9: a

Name: _______________________________

1+9=10	2+9=11	9+9=18	8+9=17
4+9=13	10+9=19	3+9=12	5+9=14
7+9=16	6+9=15	8+9=17	4+9=13
3+9=12	5+9=14	10+9=19	7+9=16
6+9=15	9+9=18	1+9=10	2+9=11
8+9=17	4+9=13	2+9=11	1+9=10
2+9=11	7+9=16	8+9=17	3+9=12
10+9=19	3+9=12	6+9=15	5+9=14
5+9=14	8+9=17	3+9=12	6+9=15
1+9=10	10+9=19	2+9=11	4+9=13
8+9=17	4+9=13	7+9=16	2+9=11

Lesson 76*   Plus 9s

Name: _______________________

4+9=13	2+9=11	9+9 =18	9+4=13
1+9=10	10+9=19	9+3=12	9+90=99
5+9=14	6+9=15	9+6=15	9+8=17
3+9=12	5+9=14	9+5=14	3+9=12
6+9=15	9+9=18	9+2=11	9+1=10
8+9=17	4+9=13	9+7=16	9+10=19
2+9=11	7+9=16	9+4=13	9+5=14
10+9=19	3+9=12	9+6=15	9+3=12
5+9=14	8+9=17	9+7=16	9+8=17
1+9=10	10+9=19	3+9=12	4+9=13
2+9=11	9+3=12	6+9=15	9+1=10

Lesson 77   Plus 9s

Name: _______________________

9+4=13   9+3=12   6+9=15   9+9=18

9+8=17   2+9=11   9+6=15   9+10=19

9+9=18   5+9=14   9+9=18   9+3=12

5+9=14   7+9=16   9+4=13   9+2=11

9+9=18   30+9=39   9+7=16   9+6=15

6+9=15   3+9=12   2+9=11   9+4=13

8+9=17   10+9=19   9+6=15   9+1=10

2+9=11   5+9=14   9+10=19   3+9=12

1+9=10   7+9=16   10+9=19   9+3=12

3+9=12   4+9=13   3+9=12   6+9=15

9+8=17   2+9=11   9+6=15   9+10=19

Lesson 78*   Plus 9s

Name: _______________________

9+6=15	5+9=14	9+9=18	0+9=9
9+2=11	6+9=15	4+9=13	8+9=17
9+1=10	3+9=12	10+9=19	2+9=11
9+8=17	8+9=17	9+3=12	9+4=13
9+9=18	1+9=10	9+7=16	9+6=15
9+10=19	4+9=13	9+3=12	7+9=16
8+9=17	2+9=11	9+10=19	4+9=13
3+9=12	9+9=18	9+8=17	5+9=14
2+9=11	9+1=10	9+80=89	9+4=13
8+9=17	9+6=15	10+9=19	3+9=12
9+0=9	4+9=13	3+9=12	9+7=16

Lesson 79   Plus 9s

Name: _______________________________

4+9=13	2+9=11	9+9=18	9+4=13
1+9=10	10+9=19	9+3=12	9+90=99
5+9=14	6+9=15	9+6=15	9+8=17
3+9=12	5+9=14	9+5=14	3+9=12
6+9=15	9+9=18	9+2=11	9+1=10
8+9=17	4+9=13	9+7=16	9+10=19
2+9=11	7+9=16	9+4=13	9+5=14
10+9=19	3+9=12	9+6=15	9+3=12
5+9=14	8+9=17	9+7=16	9+8=17
1+9=10	10+9=19	3+9=12	4+9=13
2+9=11	9+3=12	6+9=15	9+1=10

Lesson 80 T   Plus 9s

Name: _______________________

$9+6=15$	$5+9=14$	$9+9=18$	$0+9=9$
$9+2=11$	$6+9=15$	$4+9=13$	$8+9=17$
$9+1=10$	$3+9=12$	$10+9=19$	$2+9=11$
$9+8=17$	$8+9=17$	$9+3=12$	$9+4=13$
$9+9=18$	$1+9=10$	$9+7=16$	$9+6=15$
$9+10=19$	$4+9=13$	$9+3=12$	$7+9=16$
$8+9=17$	$2+9=11$	$9+10=19$	$4+9=13$
$3+9=12$	$9+9=18$	$9+8=17$	$5+9=14$
$2+9=11$	$9+1=10$	$9+80=89$	$9+4=13$
$8+9=17$	$9+6=15$	$10+9=19$	$3+9=12$
$9+0=9$	$4+9=13$	$3+9=12$	$9+7=16$

Lesson 81*   Minus 10s

Name: _______________________

15-10=5	10-10=0	16-10=6	18-10=8
17-10=7	12-10=2	14-10=4	19-10=9
16-10=6	17-10=7	15-10=5	18-10=8
13-10=3	14-10=4	17-10=7	16-10=6
10-10=0	19-10=9	15-10=5	13-10=3
12-10=2	15-10=5	17-10=7	14-10=4
11-10=1	13-10=3	10-10=0	12-10=2
15-10=5	10-10=0	16-10=6	18-10=8
13-10=3	15-10=5	17-10=7	19-10=9
19-10=9	18-10=8	14-10=4	17-10=7
16-10=6	19-10=9	12-10=2	14-10=4

Facts: -0, 1, 2, 10 mix

Name: _________________________________

5-0=5	10-2=8	16-1=15	18-1=17
7-1=6	14-10=4	17-10=7	19-10=9
16-10=6	2-1=1	14-2=12	8-1=7
3-1=2	7-1=6	15-10=5	16-10=6
10-2=8	10-10=0	17-1=16	15-10=5
2-1=1	17-0=17	18-0=18	13-0=13
11-10=1	14-2=12	14-10=4	18-10=8
5-0=5	9-2=7	10-10=0	12-1=11
3-1=2	15-2=13	18-10=8	8-1=7
19-10=9	13-10=3	7-1=6	3-0=3
12-2=10	18-10=8	10-1=9	14-2=12

Lesson 83   Minus Mix: 0s, 1s, 2s, 10s

Facts: -0, 1, 2, 10 mix

Name: _______________________

19 − 10 **9**	13 − 10 **3**	7 − 1 **6**	3 − 0 **3**	15 − 10 **5**
12 − 2 **10**	18 − 10 **8**	10 − 1 **9**	14 − 2 **12**	10 − 0 **10**
15 − 10 **5**	14 − 0 **4**	16 − 10 **6**	14 − 1 **13**	8 − 2 **6**
10 − 1 **9**	5 − 2 **3**	13 − 10 **3**	17 − 10 **7**	7 − 0 **7**
7 − 1 **6**	8 − 1 **7**	14 − 10 **4**	19 − 10 **9**	6 − 2 **4**
17 − 10 **7**	19 − 10 **9**	2 − 1 **1**	10 − 10 **0**	18 − 1 **17**
18 − 10 **8**	15 − 10 **5**	3 − 1 **2**	11 − 1 **10**	5 − 2 **3**
14 − 10 **4**	4 − 0 **4**	10 − 2 **8**	17 − 1 **16**	4 − 1 **3**
10 − 10 **0**	11 − 1 **10**	15 − 0 **15**	15 − 1 **14**	13 − 10 **3**

Lesson 84 T   Minus Mix: 0s, 1s, 2s, 10s

Facts: -0, 1, 2, 10 mix

Name: _______________________

5-0=5	10-2=8	16-1=15	18-1=17
7-1=6	14-10=4	17-10=7	19-10=9
16-10=6	2-1=1	14-2=12	8-1=7
3-1=2	7-1=6	15-10=5	16-10=6
10-2=8	10-10=0	17-1=16	15-10=5
2-1=1	17-0=17	18-0=18	13-0=13
11-10=1	14-2=12	14-10=4	18-10=8
5-0=5	9-2=7	10-10=0	12-1=11
3-1=2	15-2=13	18-10=8	8-1=7
19-10=9	13-10=3	7-1=6	3-0=3
12-2=10	18-10=8	10-1=9	14-2=12

Lesson 85*   Plus 3s   Facts: 3+ (0–5): a

Name: ___________________________

3+5=8	3+0=3	3+2=5	3+4=7
3+3=6	3+1=4	3+4=7	3+5=8
3+4=7	3+0=3	3+1=4	3+3=6
3+2=5	3+3=6	3+0=3	3+4=7
3+0=3	3+1=4	3+5=8	3+2=5
3+1=4	3+2=5	3+4=7	3+0=3
3+4=7	3+3=6	3+1=4	3+4=7
3+3=6	3+5=8	3+4=7	3+1=4
3+2=5	3+4=7	3+0=3	3+3=6
3+5=8	3+3=6	3+4=7	3+1=4
3+4=7	3+5=8	3+2=5	3+3=6

Lesson 86*   Plus 3s

Facts: 3+: b

Name: _______________________

3+6=9	3+1=4	3+5=8	3+7=10
3+10=13	3+0=3	3+2=5	3+4=7
3+3=6	3+10=13	3+7=10	3+5=8
3+4=7	3+7=10	3+8=11	3+3=6
3+2=5	3+3=6	3+9=12	3+8=11
3+7=10	3+10=13	3+5=8	3+2=5
3+1=4	3+2=5	3+7=10	3+0=3
3+3=6	3+6=9	3+8=11	3+7=10
3+8=11	3+5=8	3+4=7	3+10=13
3+9=12	3+4=7	3+0=3	3+2=5
3+7=10	3+2=5	3+1=4	3+4=7

Lesson 87 T   Plus 3s

Name: _______________

3+5=8	3+3=6	3+7=10	3+8=11
3+4=7	3+5=8	3+9=12	3+5=8
3+6=9	3+2=5	3+10=13	3+7=10
3+3=6	3+8=11	3+3=6	3+5=8
3+8=11	3+4=7	3+9=12	3+7=10
3+6=9	3+6=9	3+3=6	3+2=5
3+10=13	3+8=11	3+7=10	3+6=9
3+2=5	3+10=13	3+5=8	3+0=3
3+4=7	3+3=6	3+6=9	3+2=5
3+5=8	3+10=13	3+7=10	3+4=7
3+6=9	3+9=12	3+3=6	3+8=11

Lesson 88*   Plus 3s

Name: _______________________

8+3=11	2+3=5	4+3=7	3+3=6
4+3=7	5+3=8	0+3=3	8+3=11
9+3=12	6+3=9	10+3=13	7+3=10
5+3=8	3+3=6	7+3=10	2+3=5
6+3=9	9+3=12	0+3=3	4+3=7
8+3=11	10+3=13	4+3=7	20+3=23
2+3=5	4+3=7	5+3=8	10+3=13
7+3=10	3+3=6	8+3=11	7+3=10
5+3=8	8+3=11	6+3=9	10+3=13
1+3=4	9+3=12	4+3=7	6+3=9
8+3=11	5+3=8	7+3=10	10+3=13

Lesson 89   Plus Mix: 1s, 2s, 3s

Facts: +1, +2, +3 mix: a

Name: _______________________

9+3=12	11+3=14	5+3=8	7+3=10
5+2=7	4+1=5	12+2=14	8+3=11
10+2=12	9+2=11	11+3=14	12+2=14
8+3=11	9+3=12	2+3=5	6+2=8
6+2=8	7+3=10	11+2=13	9+3=12
9+2=11	8+3=11	2+1=3	5+2=7
5+1=6	11+3=14	7+2=9	10+3=13
8+3=11	9+1=10	10+2=12	8+1=9
11+2=13	7+3=10	12+2=14	5+2=7
10+1=11	6+2=8	5+3=8	11+2=13
5+2=7	3+3=6	9+3=12	12+2=14

Lesson 90   Plus 9s

Name: _______________________

9+6=15	5+9=14	9+9=18	0+9=9
9+2=11	6+9=15	4+9=13	8+9=17
9+1=10	3+9=12	10+9=19	2+9=11
9+8=17	8+9=17	9+3=12	9+4=13
9+9=18	1+9=10	9+7=16	9+6=15
9+10=19	4+9=13	9+3=12	7+9=16
8+9=17	2+9=11	9+10=19	4+9=13
3+9=12	9+9=18	9+8=17	5+9=14
2+9=11	9+1=10	9+80=89	9+4=13
8+9=17	9+6=15	10+9=19	3+9=12
9+0=9	4+9=13	3+9=12	9+7=16

Lesson 91*   Minus 9s

Name: _______________________

9-9=0	15-9=6	19-9=10	17-9=8
12-9=3	16-9=7	13-9=4	14-9=5
18-9=9	10-9=1	9-9=0	13-9=4
11-9=2	15-9=6	18-9=9	10-9=1
15-9=6	14-9=5	17-9=8	18-9=9
13-9=4	17-9=8	14-9=5	11-9=2
19-9=10	12-9=3	16-9=7	15-9=6
13-9=4	16-9=7	10-9=1	13-9=4
17-9=8	15-9=6	18-9=9	9-9=0
14-9=5	11-9=2	16-9=7	15-9=6
13-9=4	14-9=5	11-9=2	18-9=9

Name: _______________________

10-9=1	16-9=7	12-9=3	15-9=6
19-9=10	14-9=5	13-9=4	16-9=7
11-9=2	13-9=4	15-9=6	12-9=3
16-9=7	15-9=6	9-9=0	17-9=8
12-9=3	13-9=4	10-9=1	9-9=0
15-9=6	11-9=2	17-9=8	12-9=3
13-9=4	18-9=9	14-9=5	16-9=7
17-9=8	19-9=10	15-9=6	13-9=4
16-9=7	13-9=4	9-9=0	18-9=9
12-9=3	16-9=7	11-9=2	13-9=4
10-9=1	13-9=4	15-9=6	19-9=10

Lesson 93   Minus 9s

Name: ___________________________

11-9=2	15-9=6	9-9=0	13-9=4
15-9=6	14-9=5	13-9=4	16-9=7
9-9=0	16-9=7	18-9=9	11-9=2
14-9=5	10-9=1	12-9=3	13-9=4
19-9=10	18-9=9	17-9=8	14-9=5
11-9=2	17-9=8	14-9=5	18-9=9
9-9=0	14-9=5	16-9=7	10-9=1
13-9=4	16-9=7	10-9=1	12-9=3
17-9=8	15-9=6	19-9=10	17-9=8
14-9=5	11-9=2	16-9=7	9-9=0
16-9=7	14-9=5	11-9=2	15-9=6

Lesson 94 T   Minus 9s

Name: ______________________

10-9=1	16-9=7	14-9=5	18-9=9
19-9=10	14-9=5	18-9=9	10-9=1
11-9=2	13-9=4	15-9=6	14-9=5
16-9=7	15-9=6	9-9=0	12-9=3
17-9=8	18-9=9	14-9=5	16-9=7
15-9=6	11-9=2	10-9=1	9-9=0
13-9=4	9-9=0	17-9=8	11-9=2
14-9=5	19-9=10	15-9=6	16-9=7
17-9=8	13-9=4	18-9=9	15-9=6
12-9=3	16-9=7	11-9=2	19-9=10
10-9=1	13-9=4	15-9=6	9-9=0

Lesson 95 Plus Mix: 1s, 2s, 3s

Name: _______________

7 + 3 **10**	5 + 2 **7**	14 + 1 **15**	7 + 3 **10**	11 + 2 **13**
2 + 3 **5**	10 + 3 **13**	7 + 2 **9**	4 + 3 **7**	4 + 1 **5**
9 + 3 **12**	8 + 2 **10**	10 + 3 **13**	9 + 2 **11**	6 + 3 **9**
7 + 3 **10**	11 + 2 **13**	9 + 3 **12**	10 + 1 **11**	10 + 3 **13**
11 + 1 **12**	5 + 3 **8**	7 + 3 **10**	6 + 2 **8**	5 + 2 **7**
9 + 2 **11**	10 + 1 **11**	5 + 2 **7**	9 + 3 **12**	3 + 3 **6**
7 + 2 **9**	11 + 3 **14**	9 + 2 **11**	8 + 3 **11**	7 + 1 **8**
7 + 3 **10**	10 + 2 **12**	11 + 3 **14**	7 + 1 **8**	8 + 1 **9**
2 + 2 **4**	3 + 1 **4**	10 + 1 **11**	9 + 2 **11**	5 + 3 **8**

Lesson 96*   9 Plus Algebra

Facts: alg 9+: a

Name: _______________________

9+[2]=11	9+[6]=15	9+[0]=9	9+[4]=13
9+[6]=15	9+[5]=14	9+[4]=13	9+[7]=16
9+[0]=9	9+[7]=16	9+[9]=18	9+[5]=14
9+[8]=17	9+[1]=10	9+[3]=12	9+[4]=13
9+[10]=19	9+[9]=18	9+[7]=16	9+[6]=15
9+[2]=11	9+[8]=17	9+[5]=14	9+[9]=18
9+[0]=9	9+[5]=14	9+[7]=16	9+[1]=10
9+[4]=13	9+[7]=16	9+[1]=10	9+[3]=12
9+[8]=17	9+[6]=15	9+[10]=19	9+[8]=17
9+[5]=14	9+[2]=11	9+[7]=16	9+[0]=9
9+[7]=16	9+[5]=14	9+[2]=11	9+[6]=15

Lesson 97*   9 Plus Algebra

Facts: alg 9+: b

Name: _______________________________

9+[1]=10	9+[7]=16	9+[8]=17	9+[9]=18
9+[10]=19	9+[5]=14	9+[9]=18	9+[1]=10
9+[2]=11	9+[4]=13	9+[6]=15	9+[5]=14
9+[7]=16	9+[6]=15	9+[0]=9	9+[10]=19
9+[3]=12	9+[4]=13	9+[5]=14	9+[7]=16
9+[6]=15	9+[3]=12	9+[1]=10	9+[0]=9
9+[4]=13	9+[0]=9	9+[8]=17	9+[2]=11
9+[5]=14	9+[10]=19	9+[6]=15	9+[7]=16
9+[8]=17	9+[4]=13	9+[9]=18	9+[6]=15
9+[3]=12	9+[7]=16	9+[2]=11	9+[10]=19
9+[1]=10	9+[4]=13	9+[6]=15	9+[0]=9

Lesson 98   Plus 9s

Name: _______________________________

9+4 =13	9+3=12	6+9=15	9+9=18
9+8=17	2+9=11	9+6=15	9+10=19
9+9=18	5+9=14	9+9=18	9+3=12
5+9=14	7+9=16	9+4=13	9+2=11
9+9=18	30+9=39	9+7=16	9+6=15
6+9=15	3+9=12	2+9=11	9+4=13
8+9=17	10+9=19	9+6=15	9+1=10
2+9=11	5+9=14	9+10=19	3+9=12
1+9=10	7+9=16	10+9=19	9+3=12
3+9=12	4+9=13	3+9=12	6+9=15
9+8=17	2+9=11	9+6=15	9+10=19

Lesson 99   Plus 3s

Name: _________________________

8+3=11	2+3=5	4+3=7	3+3=6
4+3=7	5+3=8	0+3=3	8+3=11
9+3=12	6+3=9	10+3=13	7+3=10
5+3=8	3+3=6	7+3=10	2+3=5
6+3=9	9+3=12	0+3=3	4+3=7
8+3=11	10+3=13	4+3=7	20+3=23
2+3=5	4+3=7	5+3=8	10+3=13
7+3=10	3+3=6	8+3=11	7+3=10
5+3=8	8+3=11	6+3=9	10+3=13
1+3=4	9+3=12	4+3=7	6+3=9
8+3=11	5+3=8	7+3=10	10+3=13

Lesson 100   Minus 9s

Name: _______________________

10-9=1	16-9=7	12-9=3	15-9=6
19-9=10	14-9=5	13-9=4	16-9=7
11-9=2	13-9=4	15-9=6	12-9=3
16-9=7	15-9=6	9-9=0	17-9=8
12-9=3	13-9=4	10-9=1	9-9=0
15-9=6	11-9=2	17-9=8	12-9=3
13-9=4	18-9=9	14-9=5	16-9=7
17-9=8	19-9=10	15-9=6	13-9=4
16-9=7	13-9=4	9-9=0	18-9=9
12-9=3	16-9=7	11-9=2	13-9=4
10-9=1	13-9=4	15-9=6	19-9=10

Lesson 101   Minus 9s

Name: _______________________________

10-9=1	16-9=7	14-9=5	18-9=9
19-9=10	14-9=5	18-9=9	10-9=1
11-9=2	13-9=4	15-9=6	14-9=5
16-9=7	15-9=6	9-9=0	12-9=3
17-9=8	18-9=9	14-9=5	16-9=7
15-9=6	11-9=2	10-9=1	9-9=0
13-9=4	9-9=0	17-9=8	11-9=2
14-9=5	19-9=10	15-9=6	16-9=7
17-9=8	13-9=4	18-9=9	15-9=6
12-9=3	16-9=7	11-9=2	19-9=10
10-9=1	13-9=4	15-9=6	9-9=0

Lesson 102 9 Plus Algebra

Name: _______________________

9+[2]=11   9+[6]=15   9+[0]=9   9+[4]=13

9+[6]=15   9+[5]=14   9+[4]=13   9+[7]=16

9+[0]=9   9+[7]=16   9+[9]=18   9+[5]=14

9+[8]=17   9+[1]=10   9+[3]=12   9+[4]=13

9+[10]=19   9+[9]=18   9+[7]=16   9+[6]=15

9+[2]=11   9+[8]=17   9+[5]=14   9+[9]=18

9+[0]=9   9+[5]=14   9+[7]=16   9+[1]=10

9+[4]=13   9+[7]=16   9+[1]=10   9+[3]=12

9+[8]=17   9+[6]=15   9+[10]=19   9+[8]=17

9+[5]=14   9+[2]=11   9+[7]=16   9+[0]=9

9+[7]=16   9+[5]=14   9+[2]=11   9+[6]=15

Lesson 103   Plus 9s

Name: _______________________

9+6=15	5+9=14	9+9=18	0+9=9
9+2=11	6+9=15	4+9=13	8+9=17
9+1=10	3+9=12	10+9=19	2+9=11
9+8=17	8+9=17	9+3=12	9+4=13
9+9=18	1+9=10	9+7=16	9+6=15
9+10=19	4+9=13	9+3=12	7+9=16
8+9=17	2+9=11	9+10=19	4+9=13
3+9=12	9+9=18	9+8=17	5+9=14
2+9=11	9+1=10	9+80=89	9+4=13
8+9=17	9+6=15	10+9=19	3+9=12
9+0=9	4+9=13	3+9=12	9+7=16

Name: _______________________________

9+[1]=10   9+[7]=16   9+[8]=17   9+[9]=18

9+[10]=19   9+[5]=14   9+[9]=18   9+[1]=10

9+[2]=11   9+[4]=13   9+[6]=15   9+[5]=14

9+[7]=16   9+[6]=15   9+[0]=9   9+[10]=19

9+[3]=12   9+[4]=13   9+[5]=14   9+[7]=16

9+[6]=15   9+[3]=12   9+[1]=10   9+[0]=9

9+[4]=13   9+[0]=9   9+[8]=17   9+[2]=11

9+[5]=14   9+[10]=19   9+[6]=15   9+[7]=16

9+[8]=17   9+[4]=13   9+[9]=18   9+[6]=15

9+[3]=12   9+[7]=16   9+[2]=11   9+[10]=19

9+[1]=10   9+[4]=13   9+[6]=15   9+[0]=9

Lesson 105*   Plus Doubles

Facts: N + N (1–5)

Name: _______________________

4+4=8	2+2=4	3+3=6	1+1=2
3+3=6	4+4=8	2+2=4	5+5=10
0+0=0	1+1=2	5+5=10	4+4=8
2+2=4	5+5=10	4+4=8	3+3=6
5+5=10	1+1=2	3+3=6	0+0=0
1+1=2	0+0=0	2+2=4	4+4=8
5+5=10	3+3=6	1+1=2	2+2=4
4+4=8	2+2=4	0+0=0	1+1=2
0+0=0	5+5=10	4+4=8	3+3=6
1+1=2	3+3=6	2+2=4	0+0=0
2+2=4	5+5=10	1+1=2	3+3=6

Lesson 106*   Plus Doubles

Facts: N + N (5–10)

Name: ___________________________

7+7=14   9+9=18   8+8=16   10+10=20

8+8=16   7+7=14   9+9=18   6+6=12

5+5=10   10+10=20   6+6=12   7+7=14

9+9=18   6+6=12   7+7=14   8+8=16

6+6=12   10+10=20   8+8=16   5+5=10

10+10=20   5+5=10   9+9=18   7+7=14

6+6=12   8+8=16   10+10=20   9+9=18

7+7=14   9+9=18   5+5=10   10+10=20

9+9=18   6+6=12   7+7=14   8+8=16

10+10=20   8+8=16   9+9=18   5+5=10

6+6=12   5+5=10   10+10=20   8+8=16

Lesson 107   Plus Doubles

Facts: N+N (1–10)

Name: _______________________________

4+4=8   10+10=20   9+9=18   7+7=14

3+3=6   4+4=8   2+2=4   10+10=20

6+6=12   7+7=14   10+10=20   9+9=18

2+2=4   8+8=16   9+9=18   6+6=12

7+7=14   1+1=2   8+8=16   4+4=8

10+10=20   6+6=12   7+7=14   1+1=2

8+8=16   9+9=18   6+6=12   8+8=16

1+1=2   2+2=4   4+4=8   9+9=18

6+6=12   5+5=10   1+1=2   2+2=4

5+5=10   7+7=14   9+9=18   6+6=12

9+9=18   3+3=6   2+2=4   5+5=10

Lesson 108   Plus Mix: 1s, 2s, 3s

Facts: +1, +2, +3 mix: a

Name: _______________________________

9+3=12	11+3=14	5+3=8	7+3=10
5+2=7	4+1=5	12+2=14	8+3=11
10+2=12	9+2=11	11+3=14	12+2=14
8+3=11	9+3=12	2+3=5	6+2=8
6+2=8	7+3=10	11+2=13	9+3=12
9+2=11	8+3=11	2+1=3	5+2=7
5+1=6	11+3=14	7+2=9	10+3=13
8+3=11	9+1=10	10+2=12	8+1=9
11+2=13	7+3=10	12+2=14	5+2=7
10+1=11	6+2=8	5+3=8	11+2=13
5+2=7	3+3=6	9+3=12	12+2=14

Lesson 109   9 Plus Algebra

Facts: alg 9+: b

Name: _______________

9+[1]=10   9+[7]=16   9+[8]=17   9+[9]=18

9+[10]=19   9+[5]=14   9+[9]=18   9+[1]=10

9+[2]=11   9+[4]=13   9+[6]=15   9+[5]=14

9+[7]=16   9+[6]=15   9+[0]=9   9+[10]=19

9+[3]=12   9+[4]=13   9+[5]=14   9+[7]=16

9+[6]=15   9+[3]=12   9+[1]=10   9+[0]=9

9+[4]=13   9+[0]=9   9+[8]=17   9+[2]=11

9+[5]=14   9+[10]=19   9+[6]=15   9+[7]=16

9+[8]=17   9+[4]=13   9+[9]=18   9+[6]=15

9+[3]=12   9+[7]=16   9+[2]=11   9+[10]=19

9+[1]=10   9+[4]=13   9+[6]=15   9+[0]=9

Lesson 110 T   Plus Doubles

Name: _______________________________

4+4=8	2+2=4	3+3=6	1+1=2
3+3=6	4+4=8	2+2=4	5+5=10
0+0=0	1+1=2	5+5=10	4+4=8
2+2=4	5+5=10	4+4=8	3+3=6
5+5=10	1+1=2	3+3=6	0+0=0
1+1=2	0+0=0	2+2=4	4+4=8
5+5=10	3+3=6	1+1=2	2+2=4
4+4=8	2+2=4	0+0=0	1+1=2
0+0=0	5+5=10	4+4=8	3+3=6
1+1=2	3+3=6	2+2=4	0+0=0
2+2=4	5+5=10	1+1=2	3+3=6

Lesson 111*   Minus Mix: 1s, 2s, 9s, 10s

Facts: −1, −2, −9, −10 mix

Name: ___________________________

11-9=2	12-10=2	7-2=5	5-1=4
4-2=2	8-1=7	15-10=5	12-9=3
9-9=0	18-9=9	10-10=0	9-2=7
3-1=2	6-2=4	8-1=7	13-9=4
17-2=15	19-9=10	6-2=4	11-10=1
14-1=13	5-2=3	8-1=7	10-9=1
11-9=2	10-1=9	5-2=3	7-1=6
9-9=0	8-2=6	4-1=3	10-2=8
10-9=1	15-9=6	11-2=9	12-10=2
6-1=5	11-10=1	10-2=8	5-2=3
3-1=2	12-9=3	5-1=4	9-2=7

T91

Lesson 112 **T** Minus Mix: 1s, 2s, 9s, 10s

Facts: −1,−2,−9,−10 mix

Name: _______________________

5 − 1 **4**	7 − 2 **5**	12 − 9 **3**	8 − 2 **6**	12 − 2 **10**
10 − 2 **8**	4 − 1 **3**	10 − 10 **0**	7 − 2 **5**	10 − 1 **9**
11 − 1 **10**	10 − 9 **1**	13 − 10 **3**	9 − 9 **0**	6 − 2 **4**
5 − 2 **3**	6 − 1 **5**	11 − 9 **2**	12 − 2 **10**	8 − 1 **7**
10 − 2 **8**	10 − 10 **0**	9 − 1 **8**	5 − 1 **4**	12 − 9 **3**
4 − 2 **2**	4 − 1 **3**	17 − 10 **7**	9 − 9 **0**	11 − 2 **9**
10 − 9 **1**	7 − 2 **5**	10 − 1 **9**	11 − 10 **1**	9 − 1 **8**
3 − 1 **2**	11 − 2 **9**	5 − 1 **4**	12 − 2 **10**	15 − 10 **5**
5 − 2 **3**	9 − 2 **7**	13 − 2 **11**	8 − 1 **7**	7 − 2 **5**

Lesson 113*   2-Digit-2 Plus

Name: ________________________

52+9=61   82+8=90   82+4=86   52+10=62

52+3=55   82+4=86   52+7=59   52+5=57

52+2=54   82+6=88   52+8=60   52+1=53

52+7=59   82+8=90   52+9=61   52+4=56

82+0=82   82+10=92   52+5=57   52+2=54

82+10=92   82+2=84   52+3=55   32+6=38

72+3=75   32+7=39   32+1=33   72+2=74

72+4=76   32+9=41   32+0=32   72+3=75

72+2=74   72+3=75   32+10=42   72+8=80

72+7=79   72+10=82   72+3=75   32+2=34

32+5=37   32+0=32   72+6=78   72+9=81

Lesson 114   2-Digit-2 Plus

Name: ________________________

32+8=40	32+5=37	32+4=36	82+10=92
72+1=73	72+4=76	32+0=32	82+4=86
72+5=77	32+3=35	32+7=39	82+8=90
72+4=76	32+5=37	32+9=41	82+6=88
72+6=78	32+0=32	72+10=82	82+7=89
32+3=35	32+2=34	72+3=75	82+5=87
32+8=40	32+3=35	72+9=81	52+7=59
32+6=38	32+1=33	82+3=85	52+2=54
72+10=82	32+4=36	82+7=89	52+6=58
72+2=74	32+9=41	82+5=87	92+2=94
32+4=36	32+0=32	82+9=91	92+7=99

Lesson 115   9 Plus Algebra

Name: _______________________________

9+[1]=10	9+[7]=16	9+[8]=17	9+[9]=18
9+[10]=19	9+[5]=14	9+[9]=18	9+[1]=10
9+[2]=11	9+[4]=13	9+[6]=15	9+[5]=14
9+[7]=16	9+[6]=15	9+[0]=9	9+[10]=19
9+[3]=12	9+[4]=13	9+[5]=14	9+[7]=16
9+[6]=15	9+[3]=12	9+[1]=10	9+[0]=9
9+[4]=13	9+[0]=9	9+[8]=17	9+[2]=11
9+[5]=14	9+[10]=19	9+[6]=15	9+[7]=16
9+[8]=17	9+[4]=13	9+[9]=18	9+[6]=15
9+[3]=12	9+[7]=16	9+[2]=11	9+[10]=19
9+[1]=10	9+[4]=13	9+[6]=15	9+[0]=9

Lesson 116 T   2-Digit-2 Plus

Facts: 2 digit 2+

Name: _______________________________

52+9=61	82+8=90	82+4=86	52+10=62
52+3=55	82+4=86	52+7=59	52+5=57
52+2=54	82+6=88	52+8=60	52+1=53
52+7=59	82+8=90	52+9=61	52+4=56
82+0=82	82+10=92	52+5=57	52+2=54
82+10=92	82+2=84	52+3=55	32+6=38
72+3=75	32+7=39	32+1=33	72+2=74
72+4=76	32+9=41	32+0=32	72+3=75
72+2=74	72+3=75	32+10=42	72+8=80
72+7=79	72+10=82	72+3=75	32+2=34
32+5=37	32+0=32	72+6=78	72+9=81

Lesson 117   Plus Doubles

Facts: N + N (1–5)

Name: _______________________________

4+4=8	2+2=4	3+3=6	1+1=2
3+3=6	4+4=8	2+2=4	5+5=10
0+0=0	1+1=2	5+5=10	4+4=8
2+2=4	5+5=10	4+4=8	3+3=6
5+5=10	1+1=2	3+3=6	0+0=0
1+1=2	0+0=0	2+2=4	4+4=8
5+5=10	3+3=6	1+1=2	2+2=4
4+4=8	2+2=4	0+0=0	1+1=2
0+0=0	5+5=10	4+4=8	3+3=6
1+1=2	3+3=6	2+2=4	0+0=0
2+2=4	5+5=10	1+1=2	3+3=6

Lesson 118   Plus 9s

Name: _______________________________

4+9=13	2+9=11	9+9 =18	9+4=13
1+9=10	10+9=19	9+3=12	9+90=99
5+9=14	6+9=15	9+6 =15	9+8=17
3+9=12	5+9=14	9+5=14	3+9=12
6+9=15	9+9=18	9+2=11	9+1=10
8+9=17	4+9=13	9+7=16	9+10=19
2+9=11	7+9=16	9+4=13	9+5=14
10+9=19	3+9=12	9+6=15	9+3=12
5+9=14	8+9=17	9+7=16	9+8=17
1+9=10	10+9=19	3+9=12	4+9=13
2+9=11	9+3=12	6+9=15	9+1=10

Lesson 119   9 Plus Algebra

Name: _______________________

9+[2]=11	9+[6]=15	9+[0]=9	9+[4]=13
9+[6]=15	9+[5]=14	9+[4]=13	9+[7]=16
9+[0]=9	9+[7]=16	9+[9]=18	9+[5]=14
9+[8]=17	9+[1]=10	9+[3]=12	9+[4]=13
9+[10]=19	9+[9]=18	9+[7]=16	9+[6]=15
9+[2]=11	9+[8]=17	9+[5]=14	9+[9]=18
9+[0]=9	9+[5]=14	9+[7]=16	9+[1]=10
9+[4]=13	9+[7]=16	9+[1]=10	9+[3]=12
9+[8]=17	9+[6]=15	9+[10]=19	9+[8]=17
9+[5]=14	9+[2]=11	9+[7]=16	9+[0]=9
9+[7]=16	9+[5]=14	9+[2]=11	9+[6]=15

Lesson 120   2-Digit-2 Plus

Facts: 2 digit 2+

Name: ___________________________

32+8=40   32+5=37   32+4=36   82+10=92

72+1=73   72+4=76   32+0=32   82+4=86

72+5=77   32+3=35   32+7=39   82+8=90

72+4=76   32+5=37   32+9=41   82+6=88

72+6=78   32+0=32   72+10=82   82+7=89

32+3=35   32+2=34   72+3=75   82+5=87

32+8=40   32+3=35   72+9=81   52+7=59

32+6=38   32+1=33   82+3=85   52+2=54

72+10=82   32+4=36   82+7=89   52+6=58

72+2=74   32+9=41   82+5=87   92+2=94

32+4=36   32+0=32   82+9=91   92+7=99

T100